Praise for *A Mudlarking Year*

'An absolute treasure trove of sour
round inspira

'Lara Maiklem is a phenomenon.
to an epic galivant through our p
backwater but our own valley of
D

'Evocative, beautifully written and endlessly fascinating, this is a book to get lost in, whatever the season. There is a delicious sense of anticipation and discovery throughout, and the reader is rewarded with a rich and eclectic trove of finds – from a Neolithic arrowhead to a Tudor posy ring. These seemingly disparate fragments of history unite to form a dazzling patchwork of the past' **Tracy Borman**

'As with anything Lara Maiklem writes, I read it in a hungry gulp. It is totally transporting: the mysteries of the foreshore are ever enchanting and, in her hands, tangible. She writes with such elegance that I could read her day in, day out' **Sophie Dahl**

'Catches the tide of history at just the point where obsession and heroic discipline lifts the journal of an individual into revelation open to us all. And it begins to feel as if the Thames itself is dictating as much of the story as we are allowed to know' **Iain Sinclair**

'A delightful, thought-provoking book, and a profound meditation on the variety of human experience. It is as if the Thames is a great confessional for everyone who has ever lived and worked in London, and Lara is the high priestess who gently hears them and relates their lives to our own. The best historians – and the rarest – are those who look at the past and do not simply objectify it, recounting what happened and why, but intercede between the living and the dead, conveying meaning and understanding so we can see ourselves in the perspective of time. Lara is such a person: a gifted, sympathetic writer who can find poetic truth in the flotsam of forgotten lives' **Ian Mortimer**

'*A Mudlarking Year* is a beautiful reflection of what it means to be human. It is a connection to the past through both tangible history and the eternal cycle of nature that has always bound us. This is a creative, thoughtful book that makes me want to sink my own hands into the silt of the river' **Helen Carr**

'*A Mudlarking Year* is a book like no other: with her unequivocal talent and narrative flair, Lara Maiklem takes her readers on an enticing adventure and offers them the opportunity to encounter the lost objects of the past. Warm, personable, and thoroughly absorbing, *A Mudlarking Year* is a triumph of a page-turner and a truly inspirational read' **Nicola Tallis**

'A beautiful, meandering book, much like the river itself, where time and nature collide in the gloopy mud of the Thames. Enchanting, lyrical and historically fascinating' **Sam Heughan**

'Lara Maiklem is a natural historian and a born storyteller who sees history from a different perspective. Her books have stopped me in my tracks and shown how changeless we are in spirit. In her finds, Lara skewers our predecessors and we meet them as if face-to-face. She has quietly graduated from the university of unobserved places, stepping out as Professor of the Underbelly and expert in the secrets of the Thames foreshore' **Emma Bridgewater**

'*A Mudlarking Year* is a lovely annual retrospective into what attracts people to wander the stretches of the Thames foreshore in search of the capital's hidden history. Sun, rain or shine, the searcher's rituals are all too familiar and reminiscent of archaeological excavations and how the seasons can help or hinder in the process of artefact recovery. Ultimately this is a book about landscape and how humans can connect to special places they hold dear' **Raksha Dave**

'A truly wonderful book, full of fascinating details from start to finish. A must read for any Londoner, as you'll see the city with fresh eyes. I loved it!' **Alice Loxton**

'Lara Maiklem's beautiful diary of mudlarking is a powerful reflection on time, history and attention to the present moment. Her voice is rich and warm, inviting us to see beyond the surface of abandoned fragments of lost and shattered things to a deeper connection with the people of the past. For me, Maiklem's work is not unlike that of the therapist who seeks to understand the history of people's experience but must stand and look and not rush in too deep nor too fast' **Gwen Adshead**

'*A Mudlarking Year* is a love story between person and place that explores what it means to be human through the things we leave behind. Lara's passion flows from every page' **Rebecca Struthers**

'An evocative month-by-month account of Lara's addiction "to chasing lost and discarded objects and pulling stories of forgotten Londoners from the mud" . . . her truffling does more than connect her with long forgotten lives – it links her with her fellow obsessives and with the seasons as they ebb and flow along with the river' *New Statesman*

'Historically fascinating, domestic, fugitive, mournful, surreal, comic, all by turns . . . [Maiklem] is a modern Everyperson, sifting through a long human history, and coming up with literary pearls, if not often actual ones' *Tablet*

'*A Mudlarking Year* is utterly brilliant. There's a wonderful balance of observation, which feels very grounded and disciplined, with understated passion . . . The book also made me think about our (my) relationship with objects in a new light, and how the material world is infused with desire and history and identity' **Bridget Collins**

'In this book the finder is as significant as the found, the story of the discovery as valuable as the history of the object itself – and Maiklem writes as eloquently about shoe soles and her collection of broken pot handles as she does about the rare sixteenth-century posey ring that she discovers glinting in the mud' *Times Literary Supplement*

'*A Mudlarking Year* is my ideal form of non-fiction; learning history effortlessly through Lara's conversational writing style . . . Above all, the book shines for Lara's love for the subject' **David Schuster**

LARA MAIKLEM moved from her family's farm to London in the 1990s and has been mudlarking along the River Thames for over twenty years. She now lives with her family on the Kent coast within easy reach of the river, which she visits as regularly as the tides permit. Her first book, *Mudlarking*, was a critically acclaimed bestseller.

Facebook/Twitter: @londonmudlark
Instagram: @london.mudlark
www.laramaiklem.com

This QR code links to photos of Lara's finds
from this seasonal year. They can be enjoyed as a
companion to this book.

A MUDLARKING YEAR

Finding Treasure in Every Season

LARA MAIKLEM

BLOOMSBURY PUBLISHING

LONDON • OXFORD • NEW YORK • NEW DELHI • SYDNEY

BLOOMSBURY PUBLISHING
Bloomsbury Publishing Plc
50 Bedford Square, London, WC1B 3DP, UK
Bloomsbury Publishing Ireland Limited,
29 Earlsfort Terrace, Dublin 2, D02 AY28, Ireland

BLOOMSBURY, BLOOMSBURY PUBLISHING and the Bloomsbury Circus logo are
trademarks of Bloomsbury Publishing Plc

First published in Great Britain 2024
This edition published 2025

Rivers and beaches are unpredictable and potentially dangerous places. Anyone venturing
onto them should familiarise themselves with the tides, risks and hazards and take
necessary safety precautions. Always obtain the permission of the landowner to search
land or property. Searching the tidal bed of the Thames (mudlarking) requires a permit
from the Port of London Authority. The minimum age for a mudlarking permit is twelve
years old. There are areas where searching is not allowed and other locations that are
subject to legal protection, such as Sites of Special Scientific Interest (SSSIs) or Scheduled
Monuments. Mudlarking on the tidal Thames is only permitted between Teddington and
the Thames Barrier. Objects of archaeological interest should be reported to the Portable
Antiquities Scheme (PAS) – a project to record archaeological finds made by the public
in England and Wales. Objects that qualify as Treasure (essentially gold and silver over
300 years old, hoards and some other finds of exceptional significance) must be reported
to the Coroner, and Wreck (material from ships, etc.) must be reported to the Receiver
of Wrecks. For more information, websites can be found at the back of this book. The
author accepts no responsibility for personal injury or loss.

A catalogue record for this book is available from the British Library

ISBN: HB: 978-1-5266-6075-6; TPB: 978-1-5266-6074-9; PB: 978-1-5266-6078-7
eBook: 978-1-5266-6076-3; ePDF: 978-1-5266-6077-0

2 4 6 8 10 9 7 5 3 1

Typeset by Newgen KnowledgeWorks Pvt. Ltd., Chennai, India
Printed and bound in Great Britain by CPI Group (UK) Ltd, Croydon CR0 4YY

MIX
Paper | Supporting
responsible forestry
FSC® C013604

To find out more about our authors and books visit www.bloomsbury.com
and sign up for our newsletters
For product safety related questions contact productsafety@bloomsbury.com

For the river, my friend, companion and confidante.

'Sweet Thames, run softly, till I end my song.'

Prothalamion, Edmund Spenser, 1596

Contents

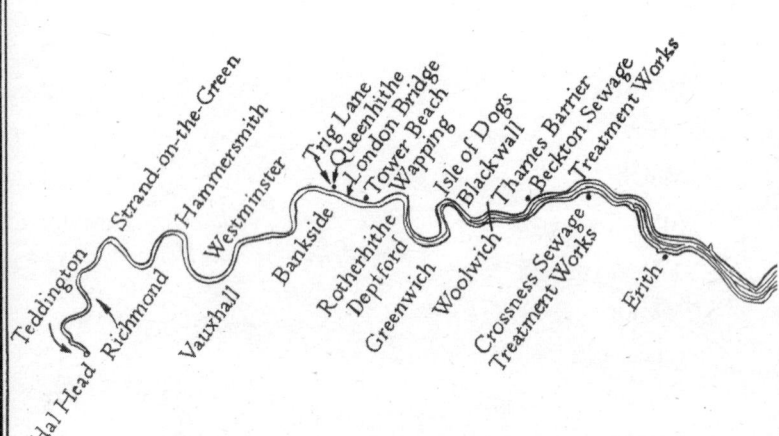

Teddington
Strand-on-the-Green
Hammersmith
Westminster
Trig Lane
Queenhithe
London Bridge
Tower Beach
Wapping
Bankside
Rotherhithe
Deptford
Greenwich
Isle of Dogs
Blackwall
Woolwich
Thames Barrier
Beckton Sewage
Treatment Works
Crossness Sewage
Treatment Works
Erith
Tidal Head
Richmond
Vauxhall

THE TIDAL THAMES

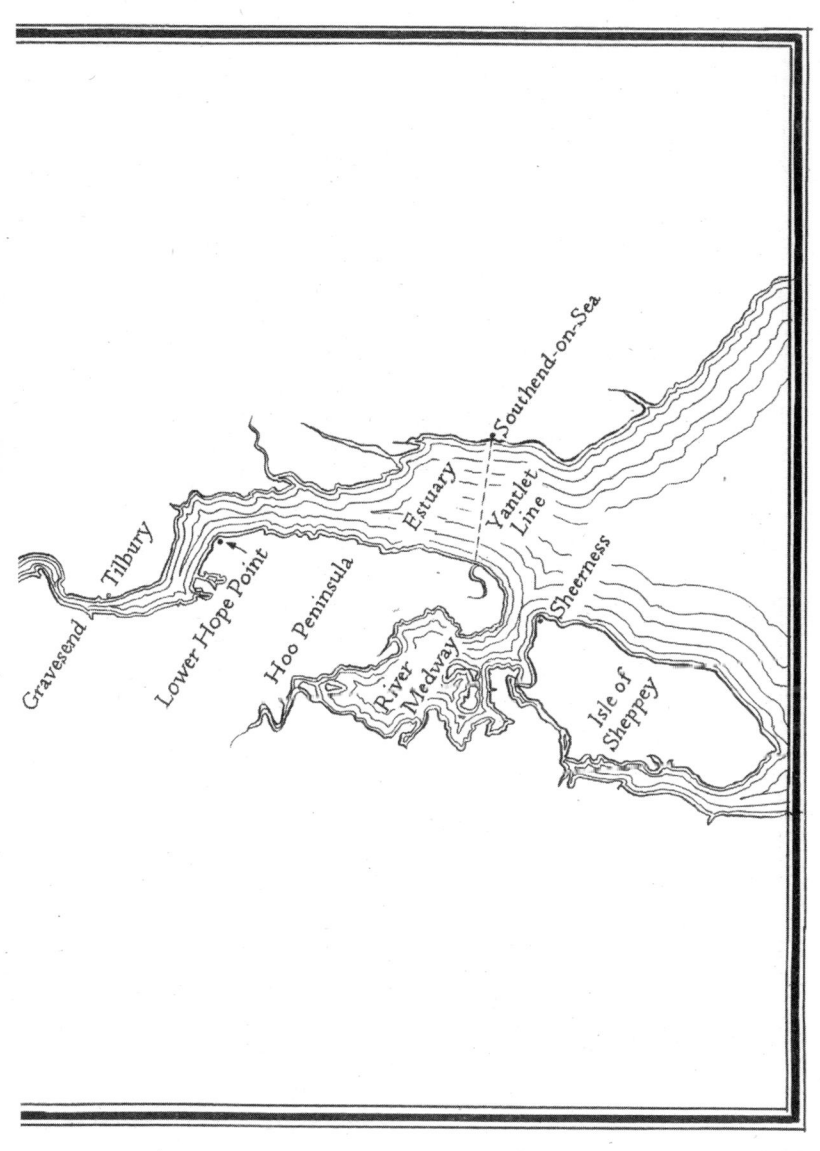

Gravesend · Tilbury · Lower Hope Point · Hoo Peninsula · River Medway · Estuary · Southend-on-Sea · Yantlet Line · Sheerness · Isle of Sheppey

WESTMINSTER

Waterloo Bridge

Blackfriars Bridge

Millennium Bridge
Trig Lane Stairs
Queenhithe Dock
Southwark Bridge

Cannon Street Bridge
London Bridge
Old Billingsgate

CITY OF
LONDON

Tower Bridge

The Town of
Wapping

Westminster Bridge

Gabriel's Wharf

Bankside

Upper Pool

Lambeth Bridge

SOUTHWARK

Lower

Vauxhall Bridge

VAUXHALL

BERMONDSEY

CENTRAL LONDON

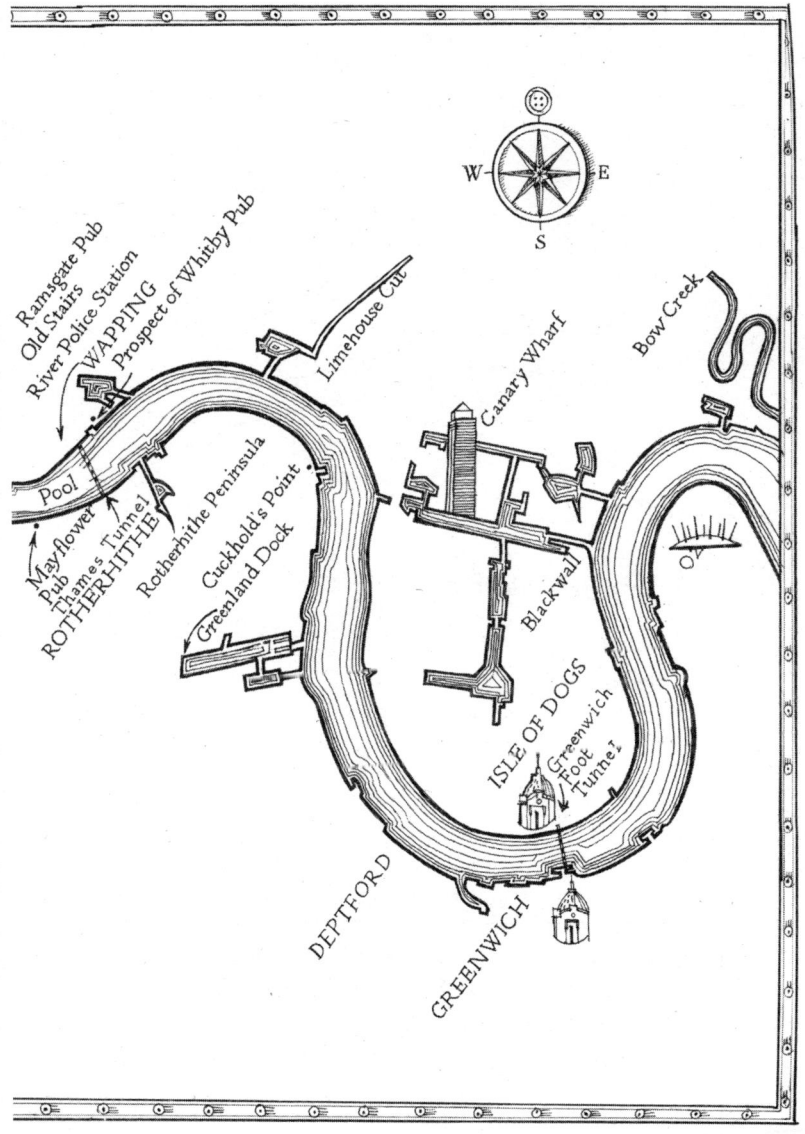

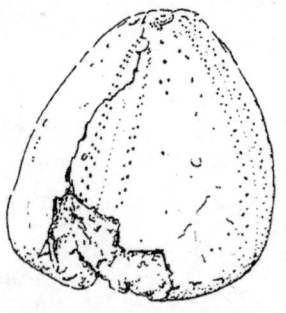

CODE: 11.22.WB.05

OBJECT:	Fossilised Sea Urchin (*Echinocorys Scutata*)
MATERIAL:	Chalk
DATE FOUND:	01/01/2022
LOCATION:	Kent
NOTES:	A complete sea urchin fossil, formed around 80 million years ago in the Cretaceous period. Old folkloric names include shepherd's crowns, fairy loaves, pixie helmets, eagle stones, chalk eggs and sugar loaves. Surface find.

Prologue

Saturday 1 January 2022 (low tide 0.69 m @ Margate, 16.46)
Kent – Margate

I'm twitchy and need to get out. I've been weighed down for days by low damp clouds, but the sun finally shines through this morning, which is a good sign. Sun on the first day of the new year.

In a perfect world I'd nip out for an hour to the Thames for a quick look at what treasures the tide had left behind on the foreshore. I could do that when I lived in London. The river was just a five-minute walk from my house in Greenwich and I'd take myself there on a whim, or if I needed escape, comfort or solitude, but we moved in 2015 and now the river is miles away. I have to plan my visits in advance, work my life around tide tables, then check train times and roadworks to make sure I can get there. What were daily visits have become less regular, but more committed, six-hour searches.

I've never lived this far from London. I grew up on a dairy farm just over twenty miles as the crow flies from central London and moved to the city in the early 1990s. In 2002 I bought a tiny, run-down Victorian two-up, two-down cottage close to the river in Greenwich. I loved my little broken house; I was never going to move, but then I was never going to have children either. The twins arrived in 2012 and while the spare room was just big enough for two cots, it wasn't big enough for two beds. So, when they were almost three, we moved, and although central London is now only seventy miles away, I am further from it than I have ever been before, and I miss it. Some days I miss it so much that it makes my heart ache. I miss the buzz, the feeling of being somewhere huge and relevant, I even miss the smell and grime, but most of all I miss the river... my river.

Mudlarking is my hobby, a comfortable obsession and not a necessity like it was for my Victorian counterparts, the half-starved scavengers, mainly women and children, who searched the foetid and sometimes frozen mud for anything they could use or sell to keep themselves alive. But while mudlarking might not be something I need to do to survive, I can't imagine life without it. I'm a time traveller and a tide traveller, addicted to chasing lost and discarded objects and pulling stories of forgotten Londoners from the mud, to watching the tides fall then rise again and the city cycle endlessly through the seasons.

Since I can't escape to the Thames today, I pile into the car with the twins, my wife Sarah and all the things she needs for swimming outside on a freezing cold day – wetsuit, gloves, bootees, float, towel, a little mat to stand on while she changes, an expensive changing robe to keep her warm, a hot-water bottle and a flask of tea. The kids throw in a football, wellies, buckets and spades, and all I bring is a canvas bag, rolled up small enough to fit in my coat pocket. I feel delightfully unencumbered.

For years, Sarah, who grew up in a landlocked part of Canada, has turned her eyes skywards and sighed in frustration at my sudden absences and obsession with weather and tides, but now she's a sea swimmer, she understands, and I'm partially off the hook. While she checks the tides at Margate I check them at London Bridge, and while she looks at water temperature and wind speed to make sure it's not too choppy, I look for rain and cold weather to make sure I'm warm and dressed to stay dry.

I hate being cold and wet. I didn't learn to swim until I was ten and I've always been scared of water. In the seven years I have lived by the sea I have never been swimming in it, and I don't intend to. Unless I can see my feet and the water is warm, I simply won't go in. My place is beside water, mooching along the tideline or poking about for curiosities, which doesn't interest Sarah, so we obsess together and separately. She swims and I lark. We both escape.

The twins are coiled springs of excess Christmas energy. They bicker and fight in the back of the car all the way to Walpole Bay tidal pool, a four-acre rectangle of trapped seawater that is reborn on every tide. It was built in 1937, the heyday of the English seaside holiday, from twelve-foot interlocking concrete blocks reinforced by reused iron tramlines. The one-ton blocks were lifted into place by hand cranes, working night and day to make use of the tides. When it was finished it was the largest tidal pool in the UK.

Occasionally the council pumps the tidal pool out completely for maintenance. This is when the metal detectorists descend like a flock of three-legged wading birds, searching for rings that have slipped off cold fingers and broken necklaces that tangle in and around the weed and lumps of chalk. I've been down there myself, but I didn't find much, just a handful of loose change.

I look out over the English Channel, which is just over sixty-five miles wide here. I am almost as close to Flanders as I am to the Thames in central London: if I'd been standing here 105 years ago, I would also have been able to hear the pounding of the guns on the Western Front. The sea between us and Europe is never clear because of the chalk that constantly erodes from the seabed and clouds the water. On a bright sunny day, it can be a beautiful verdigris, but mostly it's storm-cloud grey or even the colour of the Thames, which makes sense and makes me feel at home. I am technically only just outside the geographical boundary of the Thames Estuary, which according to the jurisdiction of the Port of London Authority (PLA) ends at Margate.

The bay is bounded by high chalk cliffs, and chalk runs beneath the yellow sand. At low tide, the sand peters out, revealing chalk that's carved into grooves and stacks by the sea and embedded with sharp grey and black flints. The same chalk forms the Thames basin, deep beneath the ground over which the river flows. It is made up of the skeletal remains of tiny planktonic green algae that floated on the upper levels of the ocean and sank down to the bottom when it died. Layers of it built up millions of years ago and slowly consolidated into chalk, which contains the fossilised remains of other creatures that lived at the same time. Today, these creatures are my quarry.

Where the sea has drilled down into the soft chalk there are rock pools and everything is covered with long, slick, brown seaweed and bright-green sea lettuce. I pick gingerly over the slippery weed, while the twins chase each other in the distance with long, dripping strands of kelp. Their squealing and laughing is the background noise to my searching on an otherwise deserted beach. I look for the smooth, egg-like domes of *Echinocorys*, sea-urchin fossils, that poke up through the sand. They are flat on the bottom and shaped like little white helmets, with patterns of dots and lines radiating from the crown. The original shell of the creature would have been covered with a fine coat of spines, which fossilised as a layer of chalky calcite, as smooth and delicate as white porcelain, and just as easily broken, cracked open and worn away. The trick is to find them before the sea does.

By the time I round up the soggy, sandy, pink-faced children, I have a heavy bag of urchin fossils to add to my collection at home, where they are lined up along windowsills and piled among flowerpots in the garden. I lug the bag back to the car – where Sarah is smiling through cold blue lips and clutching a cup of hot tea in shaking hands – and heave several million years into the boot, along with Sarah's wet swimming gear. The twins bundle into the back and Sarah folds her frozen body into the passenger seat. I turn the heater on full blast, tune out the conversation around me and think about my bench in the garage, which is vanishing under a pile of objects from the Thames foreshore that are waiting to be washed, examined and daydreamed about.

As the twins argue in the back and Sarah fiddles with the radio, I wonder if it's possible for past lives to attach themselves to objects, and I think how noisy it would be if they did. I picture the inside of the garage, the strip light flickering on and figures taking shape in the gloom. In one shadowy corner a fine bewigged Georgian gentlewoman is crying over her precious shattered blue-and-white porcelain punchbowl. I hear a noise from the wood pile and see a Victorian poet, sighing as he dips his quill into a small octagonal glass ink pot. In front of him street urchins are fighting on the floor over a penny, and as I squint across to the other side of the garage, I make out the soft form of a pregnant woman in ragged clothes who is sitting on the tumble dryer, nervously twisting a silver ring on her finger. I jump as a sailor suddenly slams his last domino down on the table among my most recent river finds. A man in plain black puritan clothes hops around him trying to pull on his left shoe, and a toothless medieval crone shouts for them all to move as she steps around the bicycles, scooters and lawnmower with a chunky brown bowl of steaming pottage in her hands. It is a babble of history that takes my mind out of the overheated fuggy car and onto the foreshore, where an invisible community of people are waiting for me.

I'm running the next few days through in my mind, trying to work out how I can get to the river, when I realise the car has gone quiet. 'Fish and chips?' says Sarah with a hint of irritation that suggests it's not the first time I've been asked. 'For dinner? Is fish and chips OK?' 'Oh yeah,' I say, pulling myself back into the here and now, 'Yeah, yeah that's fine.'

WINTER

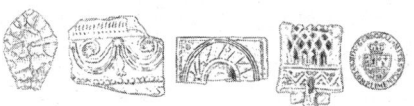

CODE: 101.22R12

OBJECT:	c.19th century toy figure
MATERIAL:	Lead
DATE FOUND:	10/01/2022
LOCATION:	Rotherhithe
NOTES:	A small (3 cm high), solid, three-dimensional lead figurine without a head or hands. Some evidence of red paint on the torso, so it may be a toy soldier. Surface find.

January

Sunday 9 January 2022 (low tide 0.86 m @ London Bridge, 12.52)
Central London – South Bank

This morning is crisp and clear, the foreshore is deserted and there is a hard granite sheen on the river's surface that draws me down onto the mud. I fumble my frozen fingers into my gloves and look across the water to the dome of St Paul's Cathedral. It is bright, white and clean in the sunshine, and every line and edge of the buildings around it are hard and precise. I make a frame with my forefingers and thumbs. The scene inside it is absolutely still, like a painting or a photograph, until a solitary seagull following the river's course west brings it to life.

It is the perfect day to bless the river, a tradition that only began in the early 2000s. The idea harks back to an ancient Orthodox Christian ceremony of throwing a cross into a body of water on the Feast of the Baptism of Jesus (the first Sunday after Epiphany), when Christians commemorate the manifestation of Christ. An epiphany is a sudden and striking realisation, and while I'm not quite expecting that, the river is the closest thing I know to a religious place, so giving thanks to its heart can only be a good way to begin the year.

It's been a while since I've mudlarked on a Sunday morning in central London, and I've forgotten how loud the city's church bells can be. The great bells of St Paul's reach me across the river and behind me Southwark Cathedral's humble answer is deep and melancholy. The lilting, trickling serenade bounces off the tall buildings of the City and rises into the bright winter sky before falling back to earth to fill streets and alleyways and flow down to the river. It sounds otherworldly, but it is just a matter of physics. Sound waves refract more in the cold, tilting and bending back down to earth when they meet warmer air, a nook of ease, higher up in the sky. This is why sound is different in the winter – louder and sharper, the intensity increased. Standing here, it is easy to imagine being on the foreshore on a cold Sunday morning in 1522, when London was a city of steeples.

If you peer between the towers of glass and steel that dominate the modern skyline of the City, you can still see some old churches, but if you want to get a really good idea of what it was like pre-Great Fire of 1666, look at Claes Visscher's panoramic map of London, which was first published in 1616. It was drawn from the south side, looking north across the river, and is a similar view to the one I captured between my fingers and thumbs. Instead of Christopher Wren's domed cathedral, Old St Paul's dominates the city, even without its 500 ft spire, which burned down after being struck by lightning in 1561, and all around it is a forest of smaller spires and bell towers.

The bells that rang from them were essential for the smooth running of London life. They called people to worship, marked the hours of the day, and were rung for festivals, celebrations, funerals and coronations. The bellringers that served each church could be called from their daily tasks at a moment's notice, but they were well compensated for their time. When Queen Elizabeth I (r. 1558–1603) was spotted travelling by royal barge, the bellringers were expected to run to their churches to salute her with a peal of bells at a charge of 3s 4d for each man.

A deep bong from the cathedral bell tower reminds me that the hours have slipped away from me again. I have wandered east as far as Southwark Bridge and I climb the slippery stone steps, away from the sixteenth century, through the heavy green iron gate at the top, and back into the crowds of tourists and day-trippers, but by the time I reach the cathedral I am too late. All that is left of the procession is the faint scent of frankincense and a verger tidying up. He points me in the direction they have gone, and I follow the smell of incense through Borough Market. I catch up with the rear of the procession on Borough High Street, not far from the site of the Tabard, the inn where Chaucer's pilgrims began their journey to Canterbury to visit the shrine of London-born archbishop St Thomas Becket, who preached his last sermon at Southwark Cathedral before leaving for Canterbury Cathedral, where he was murdered in 1170.

We are a small crowd from the south, following a line of churchmen and -women carrying croziers and wearing scarlet robes that wouldn't have looked out of place passing Chaucer's motley rabble. The original church that is now Southwark Cathedral was situated at the southern end of the medieval London Bridge, which was finished around 1209 and demolished in 1831. The church was dedicated to St Mary and became known as St Mary Overie ('over the river') for its position on the riverside. St Magnus sat opposite on the north approach and controlled two-thirds of the bridge, including the chapel, which was dedicated to St Thomas Becket and swelled the church's coffers with offerings made by pilgrims. After the Reformation, when Henry VIII (r. 1509–47) broke with Rome, the chapel on the bridge became a house, then a warehouse which was eventually demolished in the eighteenth century.

Old St Magnus burned down in the Great Fire, but the medieval bridge has left its mark on the new church that was built to replace it. When a new pedestrian walkway was added to the Old Bridge in the 1760s, a route to the bridge had to be created through the bottom of the tower at the west end of the church. When you stand under the tower today, you are directly aligned with Old London Bridge and the narrow thoroughfare, crowded buildings and chaos.

I can see the procession from St Magnus coming towards us from the north side of the bridge and we meet dead centre. A small congregation has gathered by the time the service starts and I have to strain to hear them bless not only the river, but also the people who work on it, look after it and those of us who use it for recreation. Special blessings are said for those who have died in or close to the river and my mind turns to the terrorist attack of 2017, when a van rammed pedestrians on the bridge, killing Xavier Thomas and Chrissy Archibald, and the attack in 2019, when Saskia Jones and Jack Merritt were stabbed and killed at Fishmonger's Hall. Their attacker was shot dead by police just yards from where we were standing. I'm also reminded of the day last year when an unmistakable form floated slowly past me on an ebb tide just a few feet from shore and just upstream from where the river was being blessed.

At first I thought it was a large piece of driftwood, and then I heard sirens in the distance. A bright orange RNLI dinghy, blue lights flashing, skimmed in first, and was met by a River Police boat arriving at full speed from the opposite direction. They converged on the dark floating form just a little further upstream from me and held their boats either side of it while a police officer fished about with what looked like a boat hook. Fully clothed and waterlogged, the man was limp and heavy, and it took two men to heave him out of the water onto a special stretcher clipped to the side of the police boat. They were efficient and quick, but I remember quite calmly thinking they had left something behind. The Thames is a river of lost souls and it is reluctant to release those it claims. Even after the man had been pulled free there was a sense, on that cold, quiet, grey day, that something less physical of him had been left behind to join the many others on their eternal tidal journey.

The Marine Policing Unit recovers around thirty-five bodies every year – about 90 per cent of which are attributed to suicide. Some are tragic accidents, but either way, the Thames has taken its fair share of lives over the centuries. Waterloo Bridge, a short distance but six bridges upstream from London Bridge, replaced a relatively quiet toll bridge that soon after it opened in 1817 become known as 'Lover's Leap', the 'Bridge of Sighs' and 'Arch of Suicide' for the number of people that chose to end their lives there. By the mid-nineteenth century an average of thirty people were jumping into the river from the bridge every year and it was decided to permanently moor a boat nearby to help retrieve people from the water. The boat became a floating police station and in 2002 it was sold to the RNLI for a token amount of £1. These days it gets 500 to 600 callouts a year, making it the busiest lifeboat station in the UK. Fifty per cent of their callouts are medical emergencies – heart attacks, people falling down river stairs and accidents onboard boats – the rest are people in the water.

In the nineteenth century people were paid to bring corpses ashore and attend the inquest to give evidence. This bounty led to unpleasant scenes at the site of incidents, with river workers fighting over floating bodies for the reward and the opportunity to search pockets for money, which was seen as fair spoil for their efforts. In the fourteenth century, people were less inclined to help. The *Calendar of Coroner's Rolls* recorded on 15 September 1367 a John Farnham boarding a boat at Botolph's Wharf near Billingsgate Market. While they waited for the tide to rise sufficiently to carry the boat off, a gale rose up and overturned the boat. 'His corpse was carried hither and thither until Wednesday after the feast of St Michael when it was found cast by the waters...' John Farnham lay in the water for three weeks before anyone bothered to bring him ashore.

The man I saw last September was taken east by the police boat to the temporary riverside mortuary at Wapping, a 12 ft x 12 ft metal-clad box located on a floating pontoon in front of the River Police station. Inside is a large steel tray with a drain in one end to which is attached a rubber hose for conveying leaked body juices straight into the river. Roller shutters on either side provide natural ventilation and privacy from passing boats. It is here that the bodies are 'processed': fingerprinted, searched and photographed.

A cold wind blows downriver and I turn back to the crowd. The priest's white robes flap in the breeze, incense blows away across the wet tarmac, and holy water is caught up in a sudden gust that scatters it over us like rain. A simple pine cross, much smaller than I am expecting, is brought forward and thrown off the bridge. It flutters down, lands with barely a splash and is immediately taken east by the river. I watch it bob away on dirty brown waves, another trophy for the Thames, but it is too small for this job, not large enough to soothe the ghosts that haunt the bridge and the river below.

Monday 10 January 2022 (low tide 1.07 m @ London Bridge, 13.29)
Rotherhithe

I need quiet and solitude with the river today to bid it a proper New Year greeting, so I opt for the emptiness of Rotherhithe. It takes more effort to get to Rotherhithe than it does for me to get to central London. I drive to Greenwich, take the DLR six stops, change to the Underground for one stop and the overland train for another, but it's worth it to search this long, empty stretch of the river.

The Thames is just a five-minute walk from the station. I come out, turn left and walk back on myself, heading north. I know where the river is because the sky is brighter in the distance, open and clear of buildings. I turn down a tight alleyway between a tall brick-built warehouse and a smaller, white-painted pub called the Mayflower, and I'm pleased to see the metal-barred gate at the end of the alley is open. It is at the top of a set of slippery stairs that leads down to the foreshore underneath the pub's wooden deck. It's been used by local people, river workers and sailors for centuries, but the regular tramp of muddy mudlarks' boots down the narrow alley and past the pub's entrance must annoy the landlord, and each time I visit I half expect it to be locked.

I'm finding more and more access gates locked these days. Since being caught out on an advancing tide by the sudden appearance of a generic yellow Fire Brigade padlock, I carry a set of master keys with me. This has got me out of trouble a couple of times, but I dread finding a padlock I can't open. It is illegal to put a private lock on an access gate to the river, but that doesn't stop people. Everyone wants a little bit of the river for themselves and often their way of doing that is to cut off access for others.

All along the north shore, from Limehouse through Wapping to Tower Bridge, private developers have done their best to block public access to the river, even where it is part of the Thames Path, which was established to provide a public right of way. Private estates have bent and exploited the rules with timed entry gates that only allow public access between certain hours. There are unhelpful security guards that make the process difficult, and angry signs and CCTV that make you feel as if you're breaking the rules even when you're not.

It's not just those who live beside the river who want it all for themselves though. While most mudlarks are friendly and approachable, there is avarice, territorialism and petty politics on the foreshore too. I became starkly aware of it soon after I started mudlarking, but I'm sure it's nothing new. Victorian mudlarks squabbled over territory and finds too, but these days social media has made it far easier to bully, spread rumours and say the sort of things that people probably wouldn't say in person.

I mostly managed to avoid the negative side of mudlarking until I started posting my finds online. It opened me up to abuse and I started being trolled by a handful of other mudlarks who didn't agree with my opinions about the damage mudlarking can do to the foreshore, and blamed me for telling 'other' people about their hobby.

'Troll' is a very apt name for them. Trolls lurk under bridges beside rivers, and when I'm on the foreshore I'm aware that my trolls, or those who have heard and believe the things they say, might not be too far away. Thankfully, the ones I recognise usually scuttle away quickly when they see me, studiously ignore me or glare silently as they walk past, but I have been confronted on the river too. One man walked a very long way down a completely deserted stretch of foreshore just so that he could furiously accuse me of filling it up with 'hordes of other people'.

The unlocked gate is a good omen, there are no trolls hiding under the wooden deck and Rotherhithe rewards my efforts with some good finds. A long, thick, round wooden peg, known rather eloquently as a treenail (trenail, trennel or trunnel) comes out of the mud with a satisfying slurp and the scent of ancient tar. Treenails were used in timber-framed houses, but the smell of tar suggests this once held the creaking timbers of a ship together. Rotherhithe was a graveyard for ships in the eighteenth and nineteenth centuries. They were taken here at the end of their lives to be broken up for timbers, pulled apart and cannibalised for copper. The peg may have been part of a merchant ship that had sailed around the globe, or even one of Nelson's warships, the most famous of which was the ninety-eight-gun HMS *Temeraire*, which was painted in ruins by J. M. W. Turner in 1838, being towed by steam tug to a Rotherhithe shipyard.

I collect pottery shards as I walk, and kneel to search a swathe of fine shingle. Nestled among the wet pebbles I find a single broken Georgian cufflink. It is a pewter oval with a clear glass paste 'jewel' that probably dates from around the same time as the treenail. Less than a metre from it I find a George III (r. 1760–1820) halfpenny, too worn to read the date, and a little lead figure who is fighting his way to the surface. He is just an inch high, probably Victorian, and is bravely holding up his two little fists, but without a head it's hard to say what he once was. He may be a tiny pugilist, but it's more likely he is a toy soldier that has lost the regimental colours he was carrying. I wash him in a puddle, hold him up for a better look and in the glorious absence of anyone else, I ask him out loud how he got so lost. In the bright river light and against the wide empty space he looks even smaller, and I think to myself how incredible it is, among all this, that we found each other.

Wednesday 12 January 2022 (low tide 1.44 m @ London Bridge, 15.19)
Central London – South Bank

A meeting with my publisher in town coincided with the low tide today, which would have worked out perfectly, but the meeting overran and I spent the last half-hour fidgeting under the desk, surreptitiously checking my watch and imagining the river turning and starting to rise again without me. I suppose it is a severe fear of missing out and I would say it affects most mudlarks, at least the ones I know. If I am near the river, whatever I am doing, I am usually thinking about the tide and working out how to get to it. Sometimes it's almost a relief to know the tide is high and mudlarking an impossibility and I have met people for whom it has become too much. One man had to leave the foreshore for a whole year because he was so worried about the effect his obsession was having on him. I can't imagine leaving the river though, so as soon as I am free, I race through Bloomsbury, skirt Soho, walk through Covent Garden and march quickly downhill to the Embankment.

There is a natural decline beneath the city that you feel as you walk towards the river. I bought a topographic map of the city years ago so that I could see what I was feeling. My descent to the river today begins in pinkish-red Bloomsbury at 36 m above sea level, drops to orange (32 m) at Leicester Square, yellow (25 m) on the Strand, into indigo blue (8 m) beside the river at the foot of the stairs onto Hungerford Bridge. The final length of my race to the foreshore is along the south bank, through crowds, past skateboarders, under Waterloo Bridge and past the corner where, in the summer, sand sculptors make elaborate shapes on the foreshore from the half-moon of golden sand that catches there. Today couldn't be less summery. It is damp and cold, and the light is quickly fading.

As I get down onto the mud, the sun is low, the moon is a ghost in the sky and the river has turned. The mass of footprints tells me I am not the first to get there, which isn't surprising because I'm late, but I still scan the patches of nails and iron along the edge of the river wall, just in case they missed something. There are always a lot of handmade nails here, with little triangular heads where the blacksmith finished them off with four strokes of his hammer. An armourer once contacted me to ask if I thought it might be possible to collect enough from the foreshore to make a suit of armour from authentic medieval metal. It's an interesting idea, but I told him it would be virtually impossible to accurately date the nails since they had changed so little in style over centuries. I suspect any suit of armour made from foreshore-found nails would be a glorious hybrid muddle of time.

I walk west on sliding shingle, under Blackfriars Bridge and into the strangest kind of light I've ever seen. The best mudlarking light is often a bright January afternoon, when low light casts perfect shadows that highlight even the smallest objects. But today the falling winter sun is, for a moment, perfectly positioned to reflect light off one of the city's tall glass buildings and onto the foreshore in front of me, creating an artificial brilliance that super-highlights every stone, dip and hollow. I lose myself for some time in the glorious details revealed by the light, until the sound of metal on shingle disturbs me. An old woman, bundled up against the cold, is scraping at the foreshore and she has left an ugly trail of deep scars behind her, but my irritation at being disturbed soon turns to fascination at the enormous pair of tweezers she pulls from her plastic bucket.

I've seen people searching the foreshore in all manner of ways and with all kinds of tools. They bend from the hips, kneel in the mud and sit in one place on a rock to poke about in the shingle and sand with their fingers. I've even seen one man lying completely flat on the foreshore, prostrate as a priest at an altar. They usually bring trowels to peel back the layers, but I've also seen people with hoes, rakes, dinner knives, teaspoons, forks and, of course, spades. There are some that come equipped with large garden sieves to divide the foreshore's treasures, while others use smaller kitchen sieves and old colanders. 'Look buckets', plastic tubs or buckets with the bottom cut out and replaced with clear Plexiglass, don't work well in the Thames because the water isn't clear enough, but it doesn't stop people from trying. I have seen people using normal-sized tweezers to pick up pins, and I once met a woman using them to sort through the sand literally grain by grain, but I have never seen anyone with such a huge pair of tweezers as these. In the half-light she looks like a giant crow searching for scraps in the tideline, beadily eyeing the mud and gravel, then using the tweezers like a long beak, prodding and poking around in her scrapings.

As the day slowly turns its face to the wall, the sun falls and the moon grows brighter. The super-light reflecting off the glass building switches off and the last of the sun paints the sky pink, warming the city and the dome of St Paul's with a rose blush that doesn't match the sudden drop in temperature. I zip up my jacket, blow on my frozen fingers and walk past tweezer woman as I leave the foreshore to cross the Millennium Bridge to the north side. Three gulls with sun-pinked backs fly low along the river under the bridge beneath me and from the foreshore the Thames is a mirror, reflecting the sky as it turns from pink to orange to violet to a rich indigo blue, all the colours of my topographic map. Only a plastic bottle bobbing past in the shallows breaks the surface, sending a series of shivering rings across the reflected sky.

Monday 17 January 2022 (low tide 0.97 m at London Bridge 07.48, and 1.16 m at London Bridge 20.05)

I open the tide chart on my phone, even though I know I can't get to the river today. Low tide is too early (7.48) and too late (20.05) for me and I'm relieved; also surprised to see I'm not missing a super-low one. It is a wolf moon tonight, the first full moon of the year and the brightest. By peak illumination at 6.51 p.m., it is a brilliant flimsy disc hanging in the sky, and when I go to the window to look out for the hare that lives on the moon, it is so bright it leaves green lights dancing on my retina.

Only the moon can truly control the river. Its gravitational pull as it waxes (swells) and wanes (shrinks) on its 29.5-day cycle draws the tide up and pushes it out to sea. When I'm larking, before the sun rises in the morning and when it sets in the evening, the moon is often my only companion. Even when I'm not with the river, I know that if the rising moon is over my right shoulder, I am aligned with the course it takes through central London. The moon and the river are intrinsically linked, and I know that whatever else happens in my life, the moon will keep rising and the tides will keep turning; they are my constant, my grounding and my comfort.

Monday 24 January 2022 (low tide 0.88 m at London Bridge, 12.09)
Central London – North and South Banks

I am meeting my friend Julia on the foreshore today. I haven't seen her for years. We used to meet up regularly on the river, but her visits ended abruptly about seven years ago. People come and go like the ebb and flow of the river. Over two decades I've seen a lot of new faces, but of all the curious people that find their way down onto the foreshore, relatively few become long-timers like me. Some people satisfy their curiosity with a couple of visits, while others might return for a few years, then vanish all of a sudden.

I've seen a pattern of obsession develop in those who come back again and again that's similar to falling in love. An all-consuming honeymoon follows the first find. It's a hit you want to repeat, so you come back, finding ways to weave the tides into your life. As you learn more about what you are finding and what you could find the obsession grows and a list of dreams and desires forms that can only be satisfied by visiting the river as often as possible.

Possessiveness and jealousy creep in, and in some people it becomes a mania, but in many cases it doesn't last. Life takes over, people get older, they get sick, babies are born, jobs change and people move away from the city. For some, the cold wet days and early morning starts lose their appeal and the relationship fizzles out. Only a few develop a happy, long-term, stable marriage that becomes less intense and can endure the elements, hours of travel and weeks of finding nothing. There have been months, sometimes years when I don't get to the river as much as I would like but I know it's always there and I can come back. These days I often find myself missing the river more than the thrill of the find; perhaps it's the next stage in our relationship.

I first met Julia about ten years ago, through her online blog. She wrote beautifully about her visits to the river and took her followers on a journey of discovery through her finds. But her blog was more than just a catalogue of objects and a passion for broken pottery; it was a love story, the tale of a woman with an all-consuming passion.

It began with the chance discovery of a piece of eighteenth-century Staffordshire combed slipware: a custard-yellow shard decorated with lines of brown slip pulled into waves like the top of a Bakewell tart. It opened a Pandora's box and unleashed a rush of feelings: fervour, anticipation, disappointment and yearning. She was blinded by her enthusiasm, impassioned and entirely consumed. Midway through her blog, Julia decided to make a mosaic of the Thames from the pottery and pipes she was collecting. For eight months she documented her search for specific pieces to add to the design: blue-and-white eighteenth-century German Westerwald; clay pipe stems, which she split and cut into tiny rings; chunky earthy stoneware; blue-and-orange-patterned Japanese and Chinese porcelain; shards of creamware; Victorian transferware; medieval green-glazed borderware; and, of course, more of the Staffordshire slipware that had originally captured her heart. She used around 1,500 pieces, seven centuries of London's discarded property and virtually every type of commonly found foreshore pottery. It was her *magnum opus*, but soon after it was finished, she shocked me and all her followers by announcing the end of her relationship with the river.

When I asked her later what had happened, she said in the end it was simple: she had risen at 5 a.m. to catch a tide, pulled on her wellington boots and decided to go back to bed instead. She couldn't explain what had changed, nor the strength of her obsession, but almost overnight her need for the foreshore vanished. I stayed in touch with Julia, and then at the height of Covid, she emailed me with some bad news. She had been diagnosed with ovarian cancer and the outlook wasn't great. She asked if I was available for a lark and the answer, of course, was yes.

Lovely, feisty, smiling Julia is waiting for me on the foreshore when I get there. She looks the same, perhaps a bit tired, but it doesn't take us long to get back into our easy foreshore chatter, catching up on people we used to see by the river and speculating about where they went. She tells me about the new mosaic she is making and the pottery she needs for it, so we set off to see what we can find. Julia was never interested in kneeling and peering, or the little metal things that lurk among the shingle. She's a walker and she likes to keep moving, so we search as we walk and she tells me about the last two years.

'The biggest problem with cancer is deciding how to spend your time,' she says, pausing to look out over the river. 'This may be my best year.' So she's doing the things she loves: walking, holidaying with friends, making memories for her two boys and spending time with the youngest one making a mosaic out of the pottery she's collected. 'Nobody will want it otherwise,' she says.

Julia talks easily but I find it harder so the river is a welcome distraction, a third entity that makes the conversation less intense and easier. We scan the foreshore in amiable silence for a while, looking for pieces of eighteenth-century creamware to complete a section of the mosaic. I find a very straight, dark brown stick with a round bobble at one end that I know is a Roman hairpin made of bone. It is broken, but of the many Roman hairpins I've found, it is the longest. I also find a complete decorated eighteenth-century copper-alloy cufflink embedded in rust and crusted with gravel. I show it to Julia, and we discuss the virtues of cleaning versus leaving. I can't decide what to do, so it goes in my finds bag to think about later and we head for the river stairs.

'There isn't as much pottery as I remember, and the foreshore is washing away!' she says as we reach the steep set of steps. 'It is,' I say, and I ask if she'll come back. 'Maybe,' she says. It is a strange parting, watching her vanish into the crowds with her little bag of pottery shards.

Tuesday 25 January 2022 (low tide 0.99 m at London Bridge, 12.34)
Central London – North Bank

I step off the train and follow the masses to the ticket barrier. Among the crowd I see another mudlark. I don't know who he is, but his muddy jacket and wellington boots set him apart and I spot him instantly. I follow him out of the station at a discreet distance, and as the crowd continues along the main road, we both peel off down a quiet side street and head towards the river.

Mudlarks can be easy to spot, even away from their usual territory. Outdoor wear, wellington boots, knee pads, muddy waterproofs and rucksacks usually mark us out from the regular crowd of office workers, city dwellers and tourists. Even in 1948, according to an article in *Illustrated* magazine, mudlarks had a 'look', dressing in 'old clothes and carrying walking sticks, eyebrow tweezers and small bags, they look like a group of tramps'. What the article doesn't mention, but the accompanying photographs show, is that they were also wearing suits and ties, tweed skirts, headscarves, trilby hats and brogues!

Nineteenth-century mudlarks also had a 'look', albeit wretched, poor and ragged. The journalist and sociologist Henry Mayhew's description of the women, children and old people he saw mudlarking on the Thames in 1851 is a humbling read today:

> There did not appear to be among the whole group as many filthy cotton rags to their backs as, when stitched together, would have been sufficient to form the material of shirt. There were the remnants of jackets among them, but so begrimed and tattered that it would have been difficult to have determined either the original material or make of the garment. They waded in the mud barefoot throughout the winter and any clothes they had were stiffened up like boards with dirt of every possible description.

If they had hats and bonnets, they used them to carry their finds, much like Peggy Jones, a well-known spectacle for those crossing Blackfriars Bridge around 1805, who used her heavy-duty apron folded up around her waist as a collecting bag for the coal she scavenged from the mud.

The mudlarks were still fascinating social commentators, writers and journalists almost a century later. In 1904 travel writer Dorothy Menpes described the mudlarks as a unique community with a language of their own:

> ... it is a special mud-lark patois, and appears to be a mixture of the swear words of sailors and the slang of landsmen. There are generally fights going on among these

urchins for the possession of some treasure; and their
constant enemies are the bargees, whom they delight to
cheat and annoy by clinging to ropes and chains, thereby
getting a ride gratis, as a street boy does on a Putney bus.

We may have lost our patois, but there is still a unique tribe of river searchers on the Thames who have changed little since Peter Richey, mudlark and assistant keeper at the Museum of London, described them in an article for *The Times* in 1972. He identified various types of modern mudlark that fitted into 'Six rough and sometimes overlapping categories: the specialists, the diggers, the scrap merchants, the day-trippers, the worm-catchers and the metal-detector men.'

The scrap merchants and worm-catchers may have vanished, but the rest are fairly true to form today. The specialists are still 'patient, knowledgeable, skilful and reserved', the diggers 'work on the assumption that something will turn up in a big enough hole', and the day-trippers still come 'armed with ambitiously large polythene bags ... overturn stones with frantic enthusiasm and then, as though awarding themselves a consolation prize, remove every clay pipe within reach'. Detectorists were a new phenomenon in 1972 and got short shrift from Mr Richey: 'I can never dissociate their activities as a group from plunder. Too impatient to await natural erosion, they dig where their transistors tell them to, thereby introducing a discordant note of technological aggression.' I'd like to think I qualify as a 'specialist' now, and as I follow the other mudlark through the streets to the river I wonder what type of mudlark he might be.

Thursday 27 January 2022 (low tide 1.25 m @ London Bridge, 14.28)
Wapping

The Thames in January can be quite colourless and today it seems even more transparent than ever. Perhaps it is the bitter damp wind blowing off the water that has sucked the colour out of the day. The only flash of bright colour is a red sign on the side of an old iron barge that creaks and jerks at its mooring and the bright pink hood on an otherwise bland grey jacket belonging to the only other person on the foreshore.

I make my way slowly towards the pink dot in the distance, checking over the area that was being dug up the last time I visited by a little yellow digger that had been unloaded from a barge by crane. The foreshore has settled since then and the shingle has started to creep back over the scar it made, but where the mud is still bare, I notice lots of little shards of vivid blue-and-white pottery. Maybe it is the greyness of the day that draws me to the colour, but I collect more pieces than I usually do and soon have a timeline of blue and white in my hand ranging from seventeenth-century English-made delft to nineteenth-century industrially made transferware.

A thicker, flatter piece turns out to be a fragment of seventeenth-century Dutch delft tile with the bottom half of a figure in a frock coat and stockings painted on it. Blue-and-white wall tiles were first made around 1620 and were used around fireplaces to brighten up an otherwise dark sooty hole. They were often painted with biblical and mythological scenes and scenes from daily life – people fishing, children playing, harvest time, even people going to the toilet and being sick on their way home from the tavern – scenes that must have come alive in the flickering light of a fire.

Sticking out of the mud I also spot a wedge of rough, black volcanic stone with lines radiating out towards the wider edge. It is the second piece of Roman lava quern I've found on the foreshore, so I know what it is, but I hadn't expected to find anything Roman today. Wapping is some way out of the main Roman settlement of Londinium further west, but in 2019 the remains of a bath house dating from 230–275CE were discovered at Shadwell, just a little further east from where I am standing. This suggests there was a sizeable settlement here from around the mid-third century to the early fifth century, which may have been a port or beach market associated with Londinium.

The Romans used German lava to make quern stones for grinding coarse wholemeal flour to make bread. The stones were light enough for the Roman army to carry with them, so maybe this broken quern was brought to be used by soldiers who were stationed in the watch tower above the settlement. It may never have been used at Wapping at all, but was brought to trade and intended for the city before it fell off a market stall or from the back of a cart and broke. Lava stones gradually fell out of use as people began to use windmills and watermills to grind their corn and demanded white flour without the particles of black lava grit that wore their tooth cusps smooth. By the seventeenth century lava stones were mostly only being used for grinding malt, usually for beer.

I watch a solitary crow take a freezing bath in a puddle of water left behind by the tide, and scatter a clutch of cold, hunched grey pigeons as I walk down to the water. In the winter this part of the foreshore is often covered with a layer of dead leaves, plastic rubbish, and condoms and sanitary towels from sewage spills. It catches in a dip the river has carved beside a floating jetty that channels the floating debris behind it, throwing a dirty blanket over the foreshore as the tide falls. But the foreshore is blissfully rubbish- and leaf-free today, so I rummage and search, extracting the broken mouth of an eighteenth-century wine bottle from the mud. The irregular collar, or lip, shows it was handmade by a glassmaker on a blow pipe. They are known as 'string lips' since they were applied separately as a string of glass that reinforced the mouth and offered an anchor for the wire or string that held the cork in place.

As I stand up stiffly and pocket the old piece of glass, I notice a white dome emerging from the dull grey mud. I pick around its smoothness until my finger catches an irregularity and I pull a large, exotic-looking shell free from the mud. I wash it at the river's edge, the freezing water nipping at my fingers as it reveals a very worn shell, most likely a bulla or bubble-snail shell from the Caribbean. It may have come in ballast in the belly of a trade ship returned from the West Indies filled with spices, sugar and rum, or perhaps it was a sailor's souvenir that fell out of his canvas kit bag as he staggered on sea legs along a gangplank to shore.

Exotic shells aren't an unusual foreshore find. Small cowries are the most common, and among them money cowries, small white or yellowish bean-shaped shells with smiling slits for mouths and gold-coloured rings on their backs. In the nineteenth and early twentieth centuries they were used in schools to teach children to count, but in the eighteenth century they were part of a sinister trade.

Imported in bulk from islands and atolls in the Indian Ocean, cowries were loaded into hulls as ballast and even used like polystyrene chips to pack fragile cargo, such as Chinese porcelain. In England they were transferred to ships bound for Africa, where they were used as currency to buy ivory, gold and enslaved people. Over a large part of West Africa, they became regular currency and in 1799 the Scottish explorer Mungo Park recorded that one hundred cowries would purchase a day's provisions for himself and corn for his horse. In the mid-eighteenth century, one enslaved person could be bought for 25,000 cowrie shells.

It is too cold to kneel for long, so with a little piece of warmer climes safely stashed in my bag, I set off along the waterline, stooped low against the wind, thinking about what I had read yesterday about the mudlarks that worked this patch in the mid-nineteenth century. It had struck a chord:

Some of them are old women of the lowest grade, from fifty to sixty, who occasionally wade in the mud up to the knees. One of them may be seen beside the Thames Police office, Wapping, picking up coals in the bed of the river ... She is a robust woman, dressed in an old cotton gown, with an old straw bonnet tied round with a handkerchief, and wanders about without shoes and stockings.

Henry Mayhew, *London Labour and the London Poor* (1851)

Robust or not, I imagined wading through mud on a day like today without boots or properly warm clothes. Cold weather has never kept me away from the river. I stay when it offers no comfort, when its cold, damp breath freezes me to the core, burns my cheeks and sets my teeth chattering. I stay when my frozen feet and hands turn to useless lumps and when even my eyeballs and gums feel cold. I fling my arms around and jump up and down to stay warm. To passing boats and people in warm apartments above me I must look like a crazy lady, dancing alone on the foreshore, but today the cold defeats me.

My feet are OK. I learned years ago that leather boots, even wet leather boots, are better than ordinary wellies, but my hands are in agony. Hot breath on cold fingertips only offers a moment of respite and the woolly gloves inside the latex ones aren't enough. They scream in pain, turn completely numb, then the bones begin to ache right down the core. By the time they are this cold, only a hot cup of coffee and a warm train home will return them to pink, fizzing, swollen life, so I decide it is time to leave.

I make my way towards the stairs, pausing to pick up the delicate lid of an eighteenth-century black basaltware Wedgwood coffee- or teapot. It is missing the finial, which is likely to have been a mysterious seated female figure known as 'the sibyl', an oracle or prophetess of ancient Greece. I know other mudlarks who have found the sibyl herself, but even without her the ceramic is so fine and velvety I can't resist slipping her empty throne into my bag.

I find two more things in quick succession before I wobble off the foreshore on frozen legs, face flayed raw by the cold, in search of coffee. The first is a leather shoe that I dig free from the mud with clumsy, paralysed fingers. The back quarter and outer sole and heel are missing, and its original owner's big toe has worn a hole in the middle of the upper, but apart from that, it is a perfect man's slip-on shoe. It has a slight point and a slit in the edge of the upper to make it easier to slip on, which makes me think of the type of simple shoes that can still be bought in Moroccan bazaars, but this shoe has been soaking in the mud for many years. I wonder if it might have belonged to a foreign sailor, or an English sailor who had bought a new pair of shoes in a foreign port. Contrary to popular belief, sailors preferred to wear shoes on board to keep their feet warm and because feet that are constantly wet became almost parboiled and too tender to run up rigging lines. My last find of the day is next to the shoe; perhaps they were lost or dumped together. It is an aromatic gift from the river, another wooden treenail that again releases the warm, oily, smoky scent of tar, conjuring images of rough seas, white sails and wooden ships into my frozen imagination.

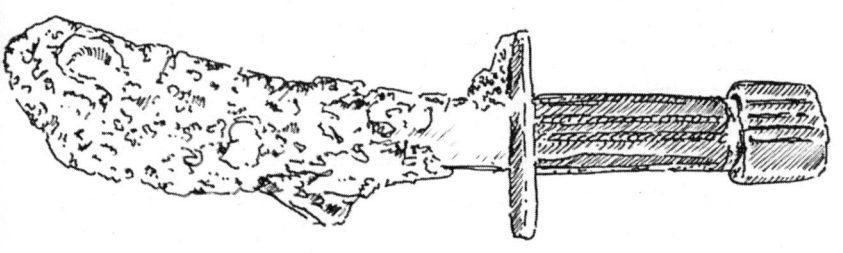

CODE: 1412.21.BB01

OBJECT:	16th century sword / bladed object
MATERIAL:	Wood, brass and iron
DATE FOUND:	14/12/2021
LOCATION:	Central London – south bank
NOTES:	A bladed object, probably a short sword. The handle is made of wood, inlaid with twisted copper alloy wire, with a square pommel finished with a quatrefoil decoration. The crossbar and blade are iron, and the blade is broken, probably in antiquity. Given to the Museum of London, uncleaned, in December 2021. Currently with the Museum's conservation department. Partially buried.

FEBRUARY

Wednesday 1 February 2022 (low tide 0.61 m @ London Bridge, 08.27)
Central London – North Bank

It's 4 a.m. and I force myself out from under a warm duvet into the cold bedroom. I dress quickly. My clothes are in a pre-prepared bundle, so I don't have to think about what to wear or rummage around for it in the dark. I pull on layer after layer – a vest, thermals, jumper, waistcoat, an extra pair of socks – and tiptoe quietly downstairs, treading to the side of the creaky stair halfway down. The kitchen is even colder than my bedroom, and the cat, fur still frosty from the garden, winds itself around my legs as I fill the kettle. I stand by the sink, squinting through my reflection in the window into the darkness, and drink my tea. The boiler fires up noisily and the radiators start to tick. They are gurgling into life as I step out into a frozen world, pulling the front door shut behind me with a soft click.

The weather on my phone promises sunshine and I hope that a few hours on the foreshore will break three days of grey dog – not quite black dog – that I haven't been able to shake. It's a melancholy that settles on me every year in January and February, the miserable months, when the land is sleeping and everything is dull and hidden, curled up tight, asleep and waiting. I hope the river will work its restorative magic and show me a glimpse of spring, a crack in the ice.

I park the car and join commuters on the 6.33 train from Greenwich to Cannon Street. Light is creeping into the sky, but there's no sign of an actual sunrise. It will be one of those mornings where the light arrives without a fanfare. By the time I descend the metal stairs by Cannon Street Bridge the tide is already quite low and it's just light enough to start searching. I turn left and head straight for London Bridge, looking towards the sky again for the sun, then I see the sign of spring I am hoping for, clinging to the river wall.

The stunted, straggly buddleia, little more than a few crispy dry twigs, that grows from an unfeasibly small crack just above the high-tide mark is showing the first signs of green at its tips. It's been there for years and although it never seems to get any larger, it produces fat purple foxtails of flowers without fail every summer. It's a beautiful survivor, nicknamed the 'bomb-site plant' because it took advantage of the city's bombscapes after World War II. I can just reach it if I stretch down from the top of the wall and occasionally, maybe twice a year, I clear its sparse branches of plastic and empty cigarette packets. Every year I expect it not to come back, but it does, a gentle symbol of resilience and a sign of hope at the end of the dark months.

Today, a small green shoot is enough to chase away grey dog, then sunlight shoulders in and fingers its way over the gravel, picking out chips of carnelian that glow as orange as Baltic amber, and quartz pebbles that shine like jewels. Actual finds are sparse – just some shards of medieval pottery, two lead musket balls, a broken seventeenth-century pewter button, a pinch of handmade dress pins and a tiny green-glass bead – but it doesn't matter. At this time of year particularly, it's not about the finds. The real treasure is chasing away grey dogs.

Thursday 2 February 2022
The Sword

I hear from the Museum of London today about the 'sixteenth-century bladed object' I found last year.

My Excalibur moment happened at the beginning of December on a murky Saturday afternoon. I don't usually mudlark on the weekend – the foreshore is often busier and I try to keep weekends for family time – but I was in London meeting friends for lunch that day, and as I made my way back to the station, I saw the tide was low. Actually, I knew the tide was low, which is why I took the longer route along the river to the station. How could I not?

I only briefly considered my new brogues, which were entirely unsuitable, before unlatching the metal gate and taking the concrete stairs down onto the foreshore at Bankside. The light was fading, and I knew I didn't have long before I lost it altogether, so I headed straight for my favourite patch. People were already there, and I could see from the footprints that it had been well searched, so I walked a little further along and that's when I saw it. I had to blink, look away and look back again to make sure I wasn't just seeing a shadow or a lucky arrangement of pebbles. It was so obvious; I couldn't believe everyone else had missed it. It was almost as if someone had planted it there, and for a moment I even considered whether they had. Was it a toy sword? Had someone left it there as a prank? But the harder I looked, the more genuine it seemed, and I began to feel light-headed.

The temptation in moments like this is to grab the find to make it real, but I resisted. This wasn't a toy or a modern replica; I could tell it was old and sensed the time that surrounded it. How old, I wasn't sure, but I knew I had one chance to savour the moment. Most finds lose something the instant they are removed from the mud that can never be replicated on a shelf or in a drawer, so I crouched down in the winter gloaming to spend time with it before I broke the spell.

The handle and hilt loomed out at me from a small area of gritty sand that thinly covered the mud into which the blade disappeared. What caught my eye was its regular shape and the two lines of twisted gold wire embedded in the dark brown material of the handle. The blade was only just beneath the surface, and I gently cleared away the sand until I felt the end of it with my fingers. Easing my hand carefully underneath, I lifted it free quite easily, leaving a perfect impression of where it had lain for the best part of 500 years in the dark grey mud. I held the sword aloft. Excalibur of the Thames! And looked around, but everyone had gone and there was nobody to share the moment with.

The handle looked to be made of wood with a square pommel carved into the end, finished off by a four-petalled flower or quatrefoil in what I assumed was copper alloy. The blade was broken at about 8 in long and was encrusted in a thick layer of mud, pebbles and rust. When iron rusts, it often engulfs whatever is lying next to it in the mud and ends up looking like a giant caddisfly larva case. If it had been a Victorian padlock or an old horseshoe, I would have been tempted to knock the concretion off with a stone, but this was too precious. The pebbly layer was protecting whatever was underneath and it needed to be preserved.

I only had a pair of latex gloves with me, just in case I ended up on the foreshore; other than that I was woefully ill-equipped. Thankfully I had picked up a copy of the *Metro* free newspaper for the train journey back, which served as some protection for the sword, and my best coat. At home, I wrapped it in wet cloths to keep it damp, sealed it in a plastic bag and quietly hid it at the back of the fridge, hoping Sarah wouldn't notice. By the following Friday it was with the conservation department at the Museum of London.

The museum confirms the handle is made of wood and the inlaid twisted wire, still as bright and shiny as the day the sword was thrown or dropped in the river, is copper alloy. The cross-bar is iron, and the conservationist thinks the blade was broken before it went in the river, so there may be an interesting story behind it. A number of daggers have been found in the Thames with deliberately broken blades. It has been suggested they were dumped by the Cutlers' Company, which had the right to confiscate substandard or foreign blades being sold in London. Perhaps this sword was one such blade, smuggled in from the Continent, confiscated by the Guild and broken on the anvil of a disgruntled swordsmith. According to the museum, the next step is to do an X-ray, which should show up a maker's mark if there is one.

I email the museum to ask how long it's going to take: 'We're very busy, and the Museum is moving this year, so it might take a while.' If the Museum wants it for its collection my problem will be solved, but if it doesn't, I will have to find someone or somewhere else to conserve it.

Thursday 3 February 2022 (low tide 0.44 m @ North Woolwich, 09.30)
The Isle of Dogs

There can be a difference of around five hours between low tide at Richmond, at the western end of the tidal Thames, and Southend in the estuary, so it is important to check the right tide table. If I am mudlarking in central London, I look at tides for London Bridge, but for Deptford, Greenwich and the Isle of Dogs I always check the tide times for North Woolwich, where low tide is usually around half an hour earlier.

I park in Greenwich, head for the small, glazed dome beside the river that is the entrance to the foot tunnel, and thank the river gods that the lift is working. The new automatic glass lifts that opened in 2012 are even less reliable than the old wood-panelled boxes that were first installed in 1904. The lifts and the tunnel replaced a ferry service that took people living south of the river to work in the docks and shipyards across the water, and for a time I also used them to get to where I worked in Canary Wharf. I know the helical staircases on both sides of the river well and have counted the steps many times: one hundred on the Greenwich south side and eighty-seven on the Isle of Dogs north side.

The old lifts were operated by men who sat all day on plastic chairs pressing the up and down buttons. For twelve hours they read their newspapers, sometimes chatted to travellers and listened to radios that fizzed and crackled until they ran out of reception 50 ft down at the bottom and midway through a song, only to buzz back into life halfway up to the top. The lifts began running at 7 a.m. and stopped precisely at 7 p.m. If you were outside those hours, you had to use the stairs, in my case with a heavy iron bicycle. But even if you were there within those hours, you weren't guaranteed a ride. The lift men's greatest glee was to close the door on cyclists who dared to cycle through the 1,215 ft long tunnel.

The white-tiled tube is always cold and damp and follows the contour of the riverbed above, sloping gently down to the middle and back up again. If I freewheeled on my bicycle, which didn't count as cycling, I could almost make it back up to the other side without having to walk and still ensure a lift up to the top. The riverbed-shaped incline and the air, which is always musty and ancient, gives a sense of not just being underground, but being under a great deal of water, which of course it is, and I enjoy the brief but delicious panic that thinking about it gives me, knowing there's a route out and clear skies at the end.

But the river did once find its way into the tunnel. At 5.30 p.m. on 7 September 1940 a bomb exploded on the foreshore on the north side of the river, twelve yards from the river wall and directly above the tunnel. It was one of the first bomb-strikes of the London Blitz and 30 ft of tile and concrete lining collapsed as a result. The river poured in and by 14 September the tunnel was completely flooded. Quick repairs were essential to keep London working, and after ten days of continuous pumping, it was finally safe to clear the debris, contain the leak and bolt thirty iron collars in place. The repair work created a section of the tunnel that is too narrow for cyclists to pass each other. It is also a permanent reminder of the power of the river above.

At the top of the north stairs, outside an identical glazed dome to the one on the opposite side, I look out across the river to Greenwich and the Old Royal Hospital, one of my favourite river views. It is how its designer Christopher Wren and his co-designer Nicholas Hawksmoor intended it to be seen, as if it were rising out of the river itself. The building is a magnificent Baroque complex of symmetrical courts, colonnades and domes, but as you stand across the river, you can also appreciate its cleverness. In the centre of this grand masterpiece, set slightly back and below a hill in Greenwich Park that's topped by the Royal Observatory, is a comparatively small white building. This is the Queen's House, which was built well before the Old Royal Hospital for King James I's (r. in England 1603–25) wife Anne. It was Queen Mary II's (r. 1689–94) request for a river view from the house that resulted in Wren's perfectly proportioned masterpiece.

I am poised at the very bottom of the Isle of Dogs and I need to make a decision. If I head west, I could probably make it to Limehouse and past the site where the *Great Eastern*, the largest steamship of its day, was launched sideways – because of its immense size – in 1858, but I'd have to come off the foreshore and do some of it along the river path. If I turn east, I can walk much further without having to climb a tall ladder to avoid pinch points. I turn east.

My access down to the mud is Newcastle Draw Dock, a wide concrete slipway within a rectangle cut into the river wall. It was built in the 1840s for loading, unloading and repairing boats and barges. There are still a number of these old slipways around the Isle of Dogs, but these days they are little more than rubbish traps, where floating driftwood and plastic piles up in nasty multicoloured jumbles. The ghosts of the old boatyards and ironworks are still trapped in the foreshore, though. Rusting chains, bolts, rivets and anonymous concreted lumps of metal gently disintegrate alongside the occasional leather boot, tools, brooms and the springs from an old mattress. It is so quiet and deserted now, it's hard to imagine how busy and noisy it once was.

When the area was redeveloped in the 1980s, rubble, tiles, roofing slate and bricks were dumped over the river wall, and I slither and slip over slimed piles of it to the water's edge. Turning east, I set off along the lines of rusted metal, conveniently sorted and arranged by the river's current into size and weight. The next tide will rearrange them again, covering and revealing even more. I crunch and clatter through the rust and metal, my eyes sweeping it for perfect circles. I find 5p, 2p, then an Edward VII (r. 1901–10) halfpenny and a George V (r. 1910–36) halfpenny (both too worn to read the dates), a small link from a gold chain and a couple of nineteenth-century trouser buttons. Then I spot something nice: a small copper-alloy military button.

The monogram of George III makes it quick and easy to find and identify on Google as I kneel over it in the mud: it is an early nineteenth-century Royal Artillery button. Between 1716 and 2003, the Royal Artillery had its headquarters at Woolwich, across the river and around the bend on the south side, just east from where I am and beyond the limits of my mudlarking licence. These days mudlarks are only permitted to search the tidal Thames between Teddington and the Thames Barrier; the rest is out of bounds. The button would have been sewn onto a short blue jacket with red facings, of the type that would have seen service during the Napoleonic Wars. It is slightly folded on one side, but the shank is unbroken, which suggests it got caught and the thread that attached it to the jacket snapped.

I bag my finds, flick the modern coins back into the river for luck and carry on. On the Isle of Dogs, the age of iron has become part of the fabric of the foreshore. In places it is like shingle, several inches thick, and where the tides have ground it down, it is a fine sand of orange rust that smells of industry, hard and metallic. A Thames Clipper, London's river bus, skims past. Apart from the planes roaring into London City Airport, it is the only real sound, and it sends a mighty wave crashing onto the foreshore, almost swamping my boots. I pass old shopping trolleys mired upside down in the mud, a bicycle with no wheels and a pink scooter that had seen better days long before it was hurled into the river. From between a docker's hobnail boot and some broken Victorian crockery I pick up a mid-century black vulcanite lemonade-bottle stopper with the familiar name 'R. Whites' stamped into it.

I keep walking, eyes down, skirting suspicious-looking patches of smooth mud that look innocuous but which I know could be deep. I clamber under a jetty and past a wide section of sandy, shingly, brick rubbly beach set into the riverside. It was once part of an eight-acre shipyard called the London Yard but it is now known as Folly House Beach and serves as the next best thing to a real beach for locals. A man and a woman are sitting on the long, shallow steps at the back end, wrapped up against the cold and engaged in an animated conversation. I must have surprised them, appearing as if out of nowhere, and they both turn to stare at me. Embarrassed at disturbing their intimacy, I look away quickly, set my eyes on the distance and speed up, only slowing down when I am past the beach and on the last stretch before the entrance to South Dock, which bars my path. This is as far as I intend to go. Any further and I will need to climb a ladder up to the river path and these days I only use ladders if I absolutely have to.

I walk slowly back, picking up the neck and shoulders of an early twentieth-century lead-and-tin toothpaste tube (Dr Matthews, London), a modern Nepalese rupee and an Indian rupee. There are no cartridges in the mud today at the spot where I've found unfired .303 World War II rounds before. There is a theory that there was a munitions factory nearby during World War II, but I've tried to research it and never found any mention of one. Maybe a crate of them fell off a barge or ship. When I find them, I don't take them home; I leave them in the mud where they are safe. The general rule on the river is to leave all ordnance alone, however large or small. Bombs from World War II and even grenades have been found by mudlarks and should be reported to the police – I've found a few anti-aircraft munitions myself. Unfired cartridges don't need to be reported but are best left where they are. Even though they are usually old, they can still have live charges and in the UK it is illegal to be in the possession of most types of ammunition without a firearms certificate.

My final find is a thick black carbon rod, about seven inches long, burnt to a taper at one end, like a giant pencil lead. It is an electric arc-lamp rod. I've found them on the Isle of Dogs before, but this one is dated, 1918. They worked in pairs by maintaining an electric arc between them. As the tips of the carbon rods heated up, the carbon vaporised and produced a very bright light. Electric arc street lamps began to appear in the 1870s and continued to be used into the early twentieth century. The light they gave off could illuminate very large areas, which wasn't always welcome. Robert Louis Stevenson, author of *Treasure Island*, described them as 'a lamp for a nightmare' compared to the 'biddable domesticated stars' provided by gaslight. Carbon rods were used in searchlights during both world wars, most famously piercing the skies over blitzed London, but the date 1918 suggests this one may have been used during World War I. Perhaps it searched the skies over the Isle of Dogs for bomber planes and the airships that switched off their engines and used the wind to float silently over the city.

Friday 4 February 2022 (low tide 0.24 m @ London Bridge, 10.51)
Central London – North Bank

A nice low spring tide today, but I almost don't go. My legs are tired after the Isle of Dogs yesterday and it is raining hard when I wake up. Sometimes it's difficult to muster the motivation, especially at 6 a.m. on a freezing cold, wet day, so I decide on an easier day in central London, not walking far but fingertip-searching a small area on the north side.

It's been throwing up some good finds for other mudlarks recently – pilgrim badges, coins, decorated medieval knife handles and even gold. When the river's being as generous as this, it's always good to be there just in case.

I am a little later than I'd like to have been, grab a coffee at the station and rustle off quickly down the road in my waterproofs towards the river. As I walk, I hope the rain might have kept other mudlarks away, but I'm wrong. The foreshore is dotted with waterproofed humps, bent over and already searching. Like me, everyone has come for the promised low spring tide. I find a spot a polite distance from anyone else, kneel down and point the tunnel of my hood at the ground. Settled and blinkered, I block out the rest of the world and begin searching.

The rain pours down in freezing sheets as I kneel next to my silent companions, peering between stones and under rubble. My rucksack bears the brunt of it, but despite two waterproof jackets and a pair of waterproof trousers, water seeps through the worn seams, soaks my jeans and trickles into my boots. Fat drops fall in a steady stream from the overhang at the top of the river wall, splattering on the rubble below and at one point it rains so hard that I can't even see across the river to the Globe Theatre on the opposite bank.

There is no escaping the elements on the foreshore unless you are near one of the bridges. You might as well be in the middle of a moor or on a clifftop for all the shelter the city offers, but that's one of the delights of mudlarking. There is nowhere to hide – you have to feel and experience the elements, seasonal changes and raw weather, while in the city people flee. They run from building to building, hide under cover and shut themselves away in temperature-controlled offices to escape the wind, the heat and the rain. Where the city paralyses and numbs the senses, the river magnifies them. The rain is wetter, the wind stronger and the sun hotter. Even the temperature drops as you descend the stairs to the foreshore. It is a microclimate within the city, a unique wilderness amid urban chaos; it is wonderful, enlivening and forgivable.

I work my way along the foreshore on my hands and knees, and people continue to come until it is too busy for me, so I move west to the stairs at Trig Lane, underneath the Millennium Bridge, where an old causeway once ran out into the river. Although the stairs are fairly modern, there has been access to the river here for centuries. Countless people have climbed into boats, filled buckets and washed their clothes here. Each journey and visit would have reaped its share of lost and discarded objects, which makes it a very good place to search.

There are usually buttons and pins, small change and modern losses from the bridge above too. Today I find £1.50 in change, a large bobble decorated with 'jewels' (glass paste) from the top of an Edwardian hatpin and a copper-alloy lock plate. It is bent, as if it had been pulled off the box or casket it was once attached to, and decorated with small, punched circles. It looks old; I take a guess at post-medieval.

I walk back through the crowds at Queenhithe, and at Cannon Street I see a familiar bike leaning against the river wall. It is Stuart's, the finds liaison officer (FLO) for London.

Stuart is responsible for recording the many objects that are found in the Thames. He works for the Portable Antiquities Scheme (PAS) as one of forty finds liaison officers across England and Wales that record the archaeological objects, mostly pre-1650, that are found by members of the public in fields, gardens, coastlines and rivers. The scheme was founded in 1997 and to date they have recorded around 1.4 million objects, which shows just how much is being found. In 2022 Stuart alone recorded around 600 objects, almost all from the Thames. The database they are recorded on is not just a unique insight into our historical lost and found, but also a valuable resource to identify finds. I use it all the time, and even though recording finds is voluntary, I'm a firm believer in contributing to it because a discovery isn't really a discovery unless it's shared.

Stuart is surrounded by mudlarks and I have to wait my turn. He's usually inundated with people wanting to know what it is they've just plucked from the mud and I feel a little guilty for bothering him, but I haven't seen him for a while, and I want to say hello. I also want to ask him about the lock plate I've just found. 'Sixteenth century – it's nice,' he confirms when I reach him at last. 'How are you?' We first started bumping into each other many years ago on remote parts of the foreshore near Wapping. That was long before he became London's FLO, when he was a simple mudlark like me.

Stuart doesn't go mudlarking any more; his visits to the river are more 'survey than searching', as he puts it. 'It's a bit of a moral dilemma because I know where everything is being found. Anyway, I love the research, it's half the lovely thing isn't it?' He's seen some amazing objects in his time as FLO, but it's still the ones with personal stories and connections that he likes the best, like the nineteenth-century docker's knife that someone brought in for him to record. 'It was so precious to its original owner that they scratched their full name into all four sides to make sure nobody stole it... only to lose it in the river.' Perhaps it's a lesson to us all not to get too attached to material objects.

Sunday 6 February 2022
Covid

I tested positive for Covid a couple of days ago. It is my first bout and thankfully it doesn't hit me too hard. Once I start feeling better, I quickly get bored of quarantine, so I decide to print out the tide tables for the next three months and set to work with a highlighter and a red pen. If I can't get to the river, then at least I can plan to get to the river.

I take each month at a time and work slowly. I'm not good with numbers – they jumble and jump on the page and get muddled in my head – so I have to concentrate hard to make sure I'm reading them right. Each dated day has two small columns, 'Time' and 'Height'. Most days have four tides, but a few only have three, where the earliest tide falls just into the previous day. I carefully mark every low tide with a small pencilled cross, then I use a highlighter to mark all the good tides under 0.5 m. With lengthening days come more daylight tides to choose from, but it is still a complicated dance between childcare, work, family life and commitments.

I am fortunate that my work is flexible. After pulling myself out of a depressing career rut twenty years ago, I began to work for myself and I have never looked back.

My multitude of ever-changing jobs and projects not only keeps my mind happy, but it also means I can fit work around life, family and of course mudlarking. While some mudlarks measure their success by the quantity, value or rarity of their finds, I measure mine simply in the amount of time I manage to get by the river. Anything else is a bonus.

I circle school holidays and cross out busy days, then I use a red pen to mark all the weekday daylight tides that I think I can get to. They need to be late enough in the day to allow for two hours of travel to get there and for three hours of searching before low tide, with another three hours after the tide turns.

Seeing the next three months in front of me, imagining the warmer weather and longer days ahead is just what I need to see off the last of grey dog.

Thursday 17 February 2022
The Society of Antiquaries

I'm on top of the world, and a bit shocked. The first email I open this morning tells me I've been elected a Fellow of the Society of Antiquaries. The election process is long and as antiquated as the society itself, which was founded on 5 December 1707 by a small group of gentlemen with antiquarian interests, who met at the Bear Tavern on the Strand in London. I was proposed by another Fellow who raised a certificate known as a Blue Paper, stating my profession and qualifications. This had to be signed by at least five other Fellows, certifying their personal knowledge of me and my worthiness. It was returned with a statement of support from my proposer to the general secretary of the society, read at an Ordinary Meeting and posted in Burlington House, the society's 'apartments' in Piccadilly, which is home to five other learned societies and the Royal Academy of Arts. My name was then put in a ballot for Fellows to vote on, requiring a ratio of two 'yes' votes to every 'no' vote. Voting was online or Fellows could vote in person on ballot days by dropping wooden balls into specially made old boxes with concealed 'YEA' and 'NO' compartments. The whole process took almost a year, but I am now Lara Maiklem, mudlark and Fellow of the Society of Antiquaries.

I knew I'd been proposed, but I never thought they would accept me, a lowly mudlark, to join the limited ranks of experts, each distinguished in their fields of archaeology, architectural or art history, and other antiquarian subjects. In the past, the closest a mudlark ever got to the society was through the objects they sold to gentlemen collectors and antiquarians, whose lives couldn't have been further removed from the squalor and misery of the scavengers.

For as long as there has been anything of value to find and people poor and desperate enough to look for it, there have probably been people searching the Thames, but it is hard to say how long these starved and pathetic creatures have been known as 'mudlarks' or how they came upon their poetic name. Perhaps it was because they were so often seen up early with the larks, working the morning tides.

In 1739 the word 'shoring' is mentioned in an Old Bailey trial record, when Edward Goynes was accused of murdering his wife Mary. Their neighbour, Catherina Lutolph, told the court that at the time of her death Mary had broken her arm while she was 'shoring'. Catherina explained that she was 'picking up what she could find upon the Shores, when the Tide was down, for Firing [to use on a fire]'.

The first written mention of the word 'mud-lark' is in *A Treatise on the Police of the Metropolis: Containing a detail of the various crimes and misdemeanours by which public and private property and security are at present injured and endangered and suggesting remedies for their prevention*, which was published in 1796. In it, the author, Scottish merchant and magistrate Patrick Colquhoun, described the gangs of criminals that were stealing from the West India merchant ships waiting to offload their precious cargoes of rum, sugar and spices. Scuffle hunters, game watermen, heavy horsemen, light horsemen, night plunderers and river pirates: each colourfully named gang had their speciality. Colquhoun estimated there to be around 10,850 thieves at work on the Thames at the time, and at the bottom of the pile were the mudlarks who lurked around the ship's hulls at low tide, looking for the packets of spices and sugar and bladders of rum that were thrown overboard for them to collect and convey to the taverns of Wapping and Rotherhithe.

The Thames River Police, the oldest continually serving police force in the world, was established on 2 July 1798 as a result of Colquhoun's book, and by the mid-nineteenth century the word 'mudlark' had come to refer to the armies of women, children and old people for whom wading through freezing cold mud was preferable to a place in the workhouse. Their quarry was anything they could use or sell: glass, rags, bones, coal, copper nails, tools that were lost by riverside workmen and, if they were lucky, objects of antiquity that they could sell to gentlemen collectors.

The mudlarks weren't the only river workers on the lookout for Bronze Age swords, medieval pilgrim badges and Roman pots to convert into cash. Over the course of the nineteenth century the river was transformed: embankments were built on both sides, Old London Bridge was torn down, new bridges and docks were built, and the channel was dredged to enable large ships to travel further upstream. Objects appeared in dredgermen's buckets, fishermen's nets and in the spoil dug out by labourers, and treasures turned up in such quantities it was said the river even vomited them up itself. According to Charles Roach Smith, antiquarian and Fellow of the Society of Antiquaries, when cofferdams were made during the building of new London Bridge in the 1830s, 'a jet of water threw up a large quantity of angels of Henry VII (r. 1485–1509) and Henry VIII, and half sovereigns of Henry VIII, which were seized directly'.

For little more than a few coins and some persistence, antiquarians could fill their cabinets and private museums with this bounty. Much of Roach Smith's collection came from the navvies rebuilding London Bridge. He collected and recorded over 5,000 antiquities and assembled them in his own Museum of London Antiquities, which he sold to the British Museum in 1856. Thomas Layton, another Fellow of the Society of Antiquaries, also built up an enviable collection of exquisite river-found objects. Layton's collection has been described as 'probably the largest collection of London antiquities ever amassed by a single individual'. 'Amassed' being an understatement. For seventy years he collected mostly from the upper stretches of the tidal Thames and obsessively hoarded. When he died in 1911 his house in Brentford and thirty outhouses were filled with boxes of antiques and antiquities, some of which, it is said, had never even been unpacked. Although he bequeathed his collection to the people of Brentford, his wish for it to form a museum in his house never came to fruition and much of it was auctioned. What was left is now scattered among various museums, libraries and the London Metropolitan Archives.

Another Fellow of the Society of Antiquaries, godfather of modern mudlarking and the man I'm proudest to follow, is Ivor Noël Hume, who described himself as an 'accidental archaeologist'. Hume was an aspiring playwright, penniless and jobless, when in 1949 he heard a man called Robin Green talking on the radio about the objects he had found on the Thames foreshore. Green was an auxiliary fireman during World War II who had fallen into the Thames in December 1940 and emerged clutching an eighteenth-century clay tobacco pipe. He described the coins, jewellery, knives and other relics he'd found scouring the foreshore, and Hume was transfixed. Since he was unemployed, he had plenty of time and began to visit the foreshore at Queenhithe. There, he found a riverside unchanged since the time of Dickens:

> *Where the warehouses along Upper Thames Street could still swing their wooden cranes out over the river to off-load bales of fur from the Hudson's Bay, chests of tea from India, cotton from Egypt and nutmeg and mace from the Molucca islands, the exotic aroma of the spices sweetening the stench of river mud.*
>
> Ivor Noël Hume, *A Passion for the Past: The Odyssey of a Transatlantic Archaeologist* (2010)

Hume soon found himself visiting up to four times a week and extending his search to the south bank between London Bridge and Southwark Bridge. His treasures included marbles, pilgrim badges, post-medieval children's toys, coins, pottery shards, buttons, pipes, wig curlers and eighteenth-century glass bottle seals, all of which would be familiar to a mudlark of today. What he couldn't identify he took to the Guildhall Museum, precursor to the Museum of London.

The curator of the museum took a liking to Hume and engaged him as a volunteer, then eventually as a paid assistant, doing what he could to help record the wealth of archaeology that was being unearthed in the hasty post-war rebuilding of London. Such was the amount that was being discovered and the lack of time and money available to recover and record it properly, the exhausted Hume, in his mud-caked duffel coat, would sometimes have to carry coal sacks filled with dripping artefacts back to the museum on the Tube.

In his time, Hume was accused of being 'just a popular archaeologist, not a professional', but it was his non-academic past and his passion for collecting that enabled him to see the worth of each object, no matter how old, commonplace or broken. Mudlarking led him to a thirty-five-year career in archaeology and a move to Colonial Williamsburg in America in 1957, where he became the director of its department of archaeology. He passed on to the great museum in the sky in 2017 and I wish above all wishes that I had met him.

Friday 18 February 2022
Storm Eunice

In November 1703, the worst storm in British history passed over London wreaking havoc and destruction. Huge waves on the Thames sent water six feet higher than ever recorded, destroying riverside homes and piling around 700 ships in the Pool of London into a tangled heap. It is now believed to have been a category two hurricane, an extra tropical cyclone, which makes our more recent storms seem tame in comparison.

Storms and wind are good for mudlarking. They churn up the river, and a good westerly can push a tide far beyond what is predicted. On Sunday 9 February 2020, just before the country went into lockdown, Storm Ciara hit the capital and the tide just kept falling. Any medieval Londoner would have seen this as a portent of doom, but the mudlarks were in heaven. The river briefly raised her skirts above her knees and all manner of treasures were gathered up that day. The most bountiful objects were old clay pipes that lay in the mud in quantities I hadn't seen since the year of abnormally low tides. People were collecting them by the fistful and one person later claimed they took over 150 of them home, which is more than I have in my entire collection. I don't know why they needed so many identical pipes – perhaps they had a reason – but I think sometimes people take things just because they are there for the taking. It's 'free stuff' they can't resist, and the fear of leaving it gets the better of them. I leave things behind or take them back to the river because I simply don't need loads of stuff and I firmly believe that in doing so I'll be rewarded on the next tide.

Today, Storm Eunice blows in hot on the heels of Storm Dudley two days ago, bringing with it a phenomenon known as a sting jet – a storm with a core of very strong winds. Eunice is the worst storm we've had in decades, and it is colliding with the spring tides. It should be a 'perfect storm' for mudlarking and I'm aching to be there for it, but unlike the Great Storm of 1703 we have had advance warning of its arrival. Trains are cancelled, the London Eye and the city's parks are closed, there is a 'red' weather warning across the whole of southern England and we have been told not to leave home unless it's absolutely necessary.

I wrestle with my conscience: the river is calling me. I could drive to it, I could risk it, I even get my rucksack out ready to go, then Sarah puts her foot down. 'You're insane – this is ridiculous. You're not going anywhere today. Put it away and do something else.' So I sit idly at my desk watching leaves barrelling across the lawn and TV aerials bending in the wind, wondering what Eunice has uncovered and what treasures I am missing.

Tuesday 22 February 2022 (low tide 0.85 m @ London Bridge, 11.45)
Central London – North Bank

The storms have almost blown themselves out, but squally gusts are taking the tops off waves and blowing them back at me as a fine, cold mist. The tides have been much lower than predicted since the storms, and today is no exception. The water flows out for over an hour longer than it should have done, and it falls much lower than the 0.85 m predicted on the tide chart. It's perfect for searching my sweet spot near the Millennium Bridge.

This small patch, less than a metre square, is only generous on very low tides and even then it is elusive. Sometimes it moves and I can't find it, then it comes back again at the whim of the river. I've snatched scores of old buttons and coins from it: a French coin dated 1643 with four mysterious drilled holes; a contemporary copy of a William III (r. 1689–1702) silver shilling, repurposed as a button; and a rather clumsy counterfeit James I silver sixpence. It might have passed as a real sixpence in a dark or a candlelit tavern, but nobody would have wanted to be caught passing a forgery. A whipping, losing an ear or a hand, branding, the stocks, dunking, fining or imprisonment? Better to chuck it in the Thames than get caught with it.

The sweet spot doesn't disappoint today. It gives me a fifteenth-century bone knife handle with a curled flourish at the end, an eighteenth-century pewter button with a weave pattern and maker's initials 'LL' on the back, four Victorian trouser buttons and a tiny seventeenth-century copper coin known as a rose farthing for the rose on one side. I regift over 30p in modern change to the river and quite a lot of foreign tourist coins. I am also rewarded with a clutch of eighteenth-century pipes.

I usually have a clay pipe or two rattling around in my mudlarking jacket pocket and it's a rare visit when I don't find one. I've pulled long and beautifully complete pipes out of soft gloopy mud and collected them by the score from tidelines. I've found them on just about every part of the foreshore, from the tidal head at Teddington to the estuary, as well as in fields, on beaches and even in my own garden. It would be impossible for even the newest mudlark not to spot the short white tubes of broken pipe stems that litter the foreshore in central London. Besides pottery shards, they are probably the most commonly found historical artefact in the UK, but that doesn't make them any less interesting.

The pipe bowl is what most mudlarks seek and if you are lucky, a bowl with a good length of unbroken stem. The bowls are very easy to date. My oldest ones are not much larger than the end of my little finger and date from the end of the 1500s. In 1573 they were described as 'little ladells' for 'the taking-in of the smoke of the Indian Herbe called Tobaco', and by the early 1600s a small pipe of tobacco could be bought at the playhouses on Bankside for 3d. At the time an unskilled labourer could expect to earn around 7d a day.

Pipes with small bowls are the oldest and the rarest because tobacco had only just arrived from the New World, and they were small because it was so expensive. Queen Elizabeth I is said to have tried it and didn't like it. James I hated it so much he wrote *A Counterblaste to Tobacco* in 1604 and raised duties on it by 4,000 per cent, from 2d to 6s 8d per pound. But despite this, it still caught on and by the mid-seventeenth century almost everyone was smoking, especially in times of disease and plague, when it was thought to keep dangerous miasmas at bay.

But pipes weren't just used for smoking. In the seventeenth and eighteenth centuries there are written accounts of pipe stems being used as ammunition in firearms, for tracheostomy tubes and catheters, as drinking straws for invalids, to drain pus from wounds and to set the bones in broken noses. Pipe bowls were used as suction devices to encourage lactation, tiny crucibles to melt lead and alloys for counterfeiting coins and, with a pig's bladder, for blowing smoke into a patient's bowels through their anus to relieve constipation. Tobacco-smoke enemas were also thought to revive drowned people and in the eighteenth century enema kits were provided by the Royal Humane Society at locations along the Thames, just in case.

It was demand that affected the size of pipe bowls. As demand grew, more tobacco was grown in the colonies, the price of it came down, people could afford more, and the size of the bowl grew to accommodate it. There are so many discarded pipes on the foreshore simply because so many people smoked, and not because pipes were thrown away like cigarette ends. A well-used clay pipe would last from a few days to a couple of weeks before it clogged up with tar and even then, it could be put in the fire to burn out the residue. When they were finally too blocked or too short for a cool smoke and became 'nose warmers', they were thrown from riverside windows, boats and bridges, or cast into the streets to be swept up and dumped in the Thames.

The precise dating of pipe bowls is quite an exact science. Along with the size and shape of the bowl, the length of a stem (the longest of all were known as a 'yard of clay'), a pointed or flat 'heel', a slight slope to the angle of a bowl, decoration and the thickness of clay can all help to date it to within a decade or less. There are very detailed clay-pipe typography guides to help, and even more specific London-variant guides. I mostly find pipes from the seventeenth to early nineteenth centuries, the golden age of clay-pipe smoking before cigarettes took over.

The pipes I find today are all eroding from one small patch of mud, where they had come to rest around 250 years ago. The unusual thing about them is that they are all armorial pipes that had been made in intricately decorated moulds. The first is a common and patriotic design, the Prince of Wales feathers, as is the second, the royal coat of arms. The third pipe features a baby sitting astride a barrel, waving a large wine glass in one hand and a round bottle the shape of an eighteenth-century onion bottle in the other. This is Bacchus, the Roman god of wine and agriculture, who also represents pleasure and revelry. The fourth pipe is lying directly underneath the Bacchus pipe. I turn to the river to wash away the mud and try to make sense of what I can see. A tree, a man in shirtsleeves and breeches, and nine long eggs – no, skittles, he's playing skittles!

Had they been plain eighteenth-century pipes, I would have left them on the foreshore, but all four armorial pipes go straight in my finds bag for my collection. These days I only keep very long plain pipes, which are fairly rare because they break so easily, and only the most perfect seventeenth-century pipe bowls that I arrange in regimented lines in the shallow wooden drawers of a small collector's chest I bought at a flea market. I've lined each drawer with different coloured felt and they look like butterflies in a lepidopterist's box. The armorial pipes will join my other decorated examples: delicate, fluted Napoleonic-era pipes with tiny oak leaves along the mould seam and pipe bowls that mimic little baskets, Turk's heads, a leg kicking a football, and a Victorian lady hitching up her skirts and squatting over a chamber pot. I'll clean these new ones properly when I get back home and rub a pinch of pencil-lead shavings over them to bring out the details.

I've found the best way to clean clay pipes is to let the mud inside dry. It shrinks away from the side of the bowl and usually comes out as a neat, dry cone, and sometimes there is a peck of unsmoked tobacco at the tip, where the embers hadn't quite reached it. I'd read about the preserved tobacco in one of Ivor Noël Hume's books some time ago, and one year I decided to gather and dry enough to smell it properly when it burned.

For months I picked shreds of tobacco out of dry river mud with tweezers and found that by crushing up the dry cones and sliding them into a wide basin of water, smaller pieces of tobacco floated to the surface, which I skimmed off with a tea strainer and dried on blotting paper. From sixty-three pipes, all of which dated from the seventeenth and eighteenth centuries, I managed to extract just under a teaspoon of dusty-looking tobacco. I laid it carefully in a line on a cigarette paper and made a very thin roll-up from it. I don't smoke, never have and didn't intend to smoke the ancient blend; I just wanted to smell it, so as a trail of white smoke curled into the air, I closed my eyes and waited.

The rich woody scent of tobacco gradually found its way to my nostrils, and in an instant I was transported back to taverns and coffee houses, cock fights and bear-baiting rings, hearthside chairs and pleasure gardens. I saw a sailor on a jetty, two men burying a linen-wrapped corpse, an old toothless woman crouched in a doorway, a man in a wig reading a newspaper, a shipwright with a hammer, a blacksmith sweating over his anvil, a clergyman and a peddler. The pipes were clamped between yellow teeth, pushed into hat bands and held in grimy hands; they were crammed into wooden boxes, rattled around in leather bags, rested on oak tables and mantelpieces and laid in bundles beside a fire. It was the scent of time and it sent a shiver down my spine.

SPRING

CODE: 138.20.LB04

OBJECT:	c.15th century pilgrim badge
MATERIAL:	Pewter
DATE FOUND:	13/08/2020
LOCATION:	Central London – north bank
NOTES:	A lozenge-shaped, cast pewter medieval pilgrim badge, featuring the mitred head and shoulders of St Osmund of Salisbury. It is missing the back pin and the four supporting angels with outstretched wings that would have created a frame around the edge. St Osmund died in 1099 and was canonised in 1457. The Reformation in 1534 ended the veneration of saints, so this badge was made sometime in the intervening years. St Osmund is the patron saint of mental illness, paralysis, ruptures and toothache. His feast day is 4 December. Surface find.

March

Thursday 3 March 2022 (low tide 0.64 m @ London Bridge, 21.14)
Central London – South Bank

A friend texts me to say she has a prebooked train ticket to London that she can't use and asks if I want it. Going up and down to London is expensive, so I text her straight back with 'yes' without thinking, then I phone Sarah. 'Fiona's got a train ticket…' I start, and before I can even finish, she replies, 'I can pick the kids up from school if you want to go.'

Any absorbing hobby can be selfish and time-swallowing. Mountaineers disappear for weeks at a time, whole Saturdays are lost to following football teams, and sailing obsessives and horse owners will spend every penny they have on mooring fees and hay. If you have an all-consuming hobby, you are constantly chasing the feeling it gives you: retreat, relief and sometimes adrenalin. Your mind is focused on doing it as often as possible; I suppose it's an addiction, and you sometimes have to use other people to get what you need. With the offer of a free train ticket came the opportunity to feed my habit. I was willing to drop everything, change my plans and travel two hours to stare at freezing cold mud by the light of a head torch, but I needed someone who could facilitate it and, as always, that was Sarah.

So, mid-afternoon, with my work done and the children taken care of, I run to the station, only just catch the train in time and make it to the foreshore... to find the tide much higher than I am expecting. What's going on? I stare at it, scratch my head, frown and look out across the water, trying to judge its speed and direction, wondering if they have closed the Thames Barrier. Everything seems normal, though; it is flowing in the right direction, but the river isn't where it should be. Still puzzling, I go off to buy a coffee to wake myself up and keep my hands warm. I sit down on the river stairs and check the tides on my phone again. It is only then I realise what I've done. My tide app is still set for the Channel. If I had been standing on the beach at home it would have been perfect, but in London, over sixty miles away from the sea, I am two hours early and the river has a long way to go before I can get onto the foreshore.

I've mistimed tides before, usually when the clocks have changed, but I've never mistimed a tide this badly, and I don't often sit by the river when the tide is up. So, with nothing else to do, I settle down on the cold steps with my coffee and watch the river inch its way seawards. As the sun lowers in the sky, it turns the rippling surface of the river into a field of wind-blown corn brushing the side of a battered barge. The barge strains and creaks at its ropes and I think of how the river was once filled with these rusted creatures, moving cargoes, powered by men pulling hard on 20 ft oars called 'sweeps'.

Waiting for the water to fall is like waiting for the proverbial pot to boil. I watch logs and other large pieces of driftwood roll past in slow motion, then I start counting cormorants to pass the time, which is difficult because they keep vanishing under the liquid-gold surface and reappearing elsewhere. I count seven, which is more than usual. Usually there are only a couple and I like to try and guess where they will pop up next.

As the sun falls, the temperature drops, and I pull on my waterproof trousers to keep warm. Dusk, gloaming, evenfall, the vesper hour, the tipping point between light and dark is one of my favourite times on the river. Darkness will come after the blue hour, as the sky balances between day and night and a rich indigo inks the heavens. The blue hour is not in fact an hour at all and rarely lasts longer than thirty minutes. It is a brief moment before twilight, in the morning and the evening, when the sun is between 4° and 8° below the horizon and the blue colour spectrum is strongest. In those beautiful, magical minutes time stands still. It is a twice-daily opportunity to reflect on the past and prepare for the future.

At the furthest ends of the tidal Thames the river makes its own dark, mercurial light. Even on moonless nights it is never completely dark, and in central London it is in a perpetual state of synthetic half-light. As the blue light fades, the lights under the bridges come on. They join the stacks of lights in the City and the strings of bulbs along the riverside that reflect off the water. On the other side of the river, I can see two lights bobbing around like will-o-the-wisps. There are other nightlarks out tonight and I hope they aren't using the cover of darkness to do things they shouldn't. Recently someone tried to wrench a Saxon fish trap out of the mud at Chelsea, and cartwheels and anchors have also vanished overnight from other locations, with rumours that people are coming by boat on late low tides to steal them for garden ornaments. Some objects, especially the larger ones, belong to the river and should be left where they are.

I squint up through the light pollution, make out a few of the brightest stars, click on my own head torch and step onto the thin wet sliver of foreshore that's gradually widening at the bottom of the stairs. As the river slowly pours itself back into the sea, I start looking for the treasures it has left behind.

Friday 4 March 2022
The Finds Liaison Officer

Thanks to Covid restrictions, it has been over two years since I last visited my finds liaison officer and I have quite a pile to show him. Jo the FLO, the finds liaison officer for Kent, inhabits the corner of a depressing open-plan office at Kent County Hall in Maidstone. Since I moved to Kent it has been easier and faster for me to take my finds to Jo than to lug them up to London, where Stuart is always drowning under river finds.

Some mudlarks record their finds themselves, but I like seeing Jo. He's a gentle soul in a leather jacket, with a mohawk that's tamed for work and a tattoo of a medieval London penny on his forearm. His eyes are never still, and he has a restless energy that drives his enthusiasm for what he does. Jo is an archaeologist by training and he mostly records objects that are found in Kentish fields by metal detectorists, but his patch also includes the river Medway. We talk about the problems there, people stripping the area of finds and trespassing on Deadman's Island. Some of them are using hovercrafts that disturb protected nesting sites, and the RNLI has been called out to people trying to walk out to the islands across the mud at low tide, which is very dangerous.

As we talk, I begin to unwrap my carefully selected treasures. I know Jo will only be able to take about ten of them, so I have already whittled them down to unusual objects and things I am pretty sure are over 300 years old. Jo quickly shuffles them into two piles, 'keep' and 'reject'. There is one particular mystery: a bead shaped like a tiny pot that I had found in an area well known for Roman artefacts and pottery. All through lockdown I'd been searching reference books and raking through the internet for something to compare it to and drawn a blank, but Jo hones in on it straight away. 'That's interesting. I think I've seen something like it before, but it's not made of clay – it's metal. Leave it with me.'

I pull out a squarish shard of decorated orange Roman Samian ware that I am particularly proud of identifying myself. Actually, I've only been instrumental in identifying it because an archaeologist had seen it on my Instagram page and told me what it was. 'It's a penis dog!' I announce gleefully. Jo looks at me doubtfully. I point out what I had originally thought was a funny-looking tree, then turn it upside down and there it is, a clear phallus with two long hind legs and a tail. It wasn't easy to research – in fact, I wouldn't advise anyone to search 'penis dog' on the internet – but I had eventually discovered that Romans were very keen on phallus-shaped charms. They thought they offered protection or, failing that, were so ridiculously weird and obscene that they could send an enchanter mad. It had certainly played on my sanity since I'd found it.

The other object I am proud to have identified is a roughly made grey ceramic tube, about as wide as my upper arm and half its length. It narrows towards one end and the colour and type of clay used to make it matches the type of utilitarian Roman pottery that's commonly found on the foreshore. I was pretty sure it was Roman, so I had contacted a Roman pottery specialist, who confirmed what it was: 'A bellows nozzle or tuyére for use in a Roman furnace, probably a metal working furnace.' A wooden tube to which a pair of bellows was attached would have been fastened to the tapered end and the other would have gone straight into the furnace. The ceramic nozzle protected the bellows from direct contact with the fire so that air could be safely blown into the furnace to increase the heat. It is unusual and deserves to be recorded.

We rummage through boxes and bags for over an hour and in the end, Jo takes far more than he should. By the time we have finished chatting and I am ready to go, twenty-six of my objects are safely packed away and filed in boxes on shelves with all the other finds he has to record. 'Come back in September,' he says, 'they should be ready to collect by then.'

Tuesday 8 March 2022 (low tide 0.87 m @ London Bridge, 11.33)
Central London – North Bank

My first grey wagtail with a bright lemon-yellow belly flits up and away as I step off the stairs onto the mud today. I've only ever seen black-and-white pied wagtails on the foreshore before. He flies back towards me, lands on a pile of bricks close by, and I notice he only has one foot. London birds are often missing feet or toes. People used to say acid was put on building ledges to stop them from landing and making a mess, but I don't think that's true. Bits of thread, plastic line and other detritus catch around their legs and toes. It eventually cuts off the blood supply and they simply fall off. Somewhere in London there are little toes and feet, blown together in gutters or caught between the bricks of chimneys.

The sun is out and there is a sense of spring in the air. Pigeons are dancing along the waterline, bobbing and spinning, and puffing out their iridescent necks in elaborate courtship. These cliff-dwelling rock doves are perfectly adapted to city life. They raise their young under the bridges over the river and can produce up to eight broods a year, which they nourish with 'milk' secreted from the lining of their crop. I sometimes find their pure white eggs smashed on the foreshore and the broken transparent bodies of chicks that have fallen from their nests: tiny, reptilian, dinosaur-like things with impossibly large hearts.

I walk to a stretch where the foreshore dips into a natural bay that's been eroded by the river, and stop in the sunshine to talk to another mudlark. He points to the base of the wall: 'Look at that, it's getting worse.' And it is. Each time I visit it seems the river has picked away a bit more of the foreshore by the wall and the foundations are now completely exposed. I see why as I watch the waves wash in after a passing Clipper. They crash against the wall, bounce off it, hit another wave and wash back into the wall again. This is happening over and over again, turning the bay into a swirling twin-tub of water, worrying away the foreshore, undermining the wall and in the process churning up a trove.

Erosion is a double-edged sword that reveals as it destroys. Soon they will have to fill in this gap, or the wall will start to crack and sink. They will probably rebuild the revetment and fill it in with nets of rocks and hardcore, as they have in other places along this stretch. It will cover a fruitful spot and the regular mudlarks are dreading it.

The little crescent-moon-shaped bay has become known as Roman Bay, or the Roman hole, for the number of Roman objects that have been found here. They are thought to be washing in from further out, perhaps from a dump of spoil that was dug out during building works further into the city in the eighteenth or nineteenth century but nobody is really sure. Coins, pottery, beads, shards of ancient glass, bone hairpins and game counters, chunks of mosaic floor, even small statues of gods from household shrines, flue tiles from Roman central-heating systems and thick pieces of roof tile with the dotted imprints of studded sandal soles have all been found here over the years. Sometimes they are covered in a thin, cream-coloured crust called 'Thames race', a natural river phenomenon that takes millennia to build up and only covers the oldest of the river's hoard.

For a few years about ten years ago, Roman Bay was my only haunt, and I'd visit on every tide I could. It was a quiet spot that didn't attract the detectorists and scrapers because it was protected. Only objects lying on the surface could be collected, and the finds were good. But the ban was lifted for members of the Society of Mudlarks, a members-only group, and they started to scrape away the surface. It got busier, finds became scarcer and I stopped going so often.

Today, six other mudlarks are already silently lined up along the waterline, diligently following the tide as it falls. The trick is to walk up and down the line of water, watching the point where the waves wash against the sand. If you are lucky, the river will reveal something or drop it at your feet. The last time it did that for me I was gifted a Charles I (r. 1625–49) groat, but it has also washed in a Roman melon bead, a Charles I shilling and a beautiful post-medieval jetton. All conjured from the depths as if by magic.

Each time a boat comes close, the line steps back in unison away from the wash, then slowly edges back as it calms. Suddenly a huge wave scatters them, but one person isn't fast enough. The river crashes against his legs, soaking them to the knees, and a collective groan rises from the others in sympathy. The boats don't always send waves, though. Sometimes the water swells suddenly and silently without warning; it's a choreographed dance with the river that can catch out even the most experienced of dance partners.

I walk the line with the others, dodging the waves as they crash in. I collect some small shards of Roman pottery that were once delicate black-and-chocolate-brown beakers, rough black cooking pots and a Samian bowl that had probably been brought from Gaul, an area comprising modern-day France, parts of Belgium, western Germany and northern Italy.

I look down at the little collection of pottery in my hand and I let my mind wander away from the crowded foreshore.

A boy is running into a kitchen and grabbing a fine orange Samian bowl off a shelf. He ladles stew into the bowl with a shaking hand and returns to the main room, where his master is sitting hunched over a wooden table.

He had brought the boy with him as a house servant almost two years ago, when he arrived in Londinium from Rome. He had always been a brooding, surly man, but with every passing month he had become more morose and prone to angry outbursts. The boy knew his master hated this far-flung province. He hated the people with their strange barbaric ways, he hated the cold fogs that rolled in off the dull grey river, but most of all he hated the rain. When it rained, he drank, and it hadn't stopped raining for days. The boy crosses the room as quietly as he can, gently sets the bowl with a spoon next to his master's arm and stands back. The man sits up without a word and stares at the boy for a moment as if contemplating every bad decision that has landed him in this stinking cesspool of a city, then he sweeps the bowl onto the floor with his arm where it shatters into pieces at the boy's feet.

Monday 14 March 2022 (low tide 1.83 m @ North Woolwich, 16.42)
Greenwich

I had to go to Greenwich for a meeting today and decided to wait out the worst of the rush-hour traffic in a familiar spot before setting off home. I've been searching the river at Greenwich for over twenty years, but it's a fickle place. I tend to either find something straight away or nothing at all.

The mud mostly contains animal bones, oyster shells and broken pottery from the Tudor palace that is now little more than a pile of rubble on the foreshore and foundations below riverside lawns. By the sixteenth century, the palace had grown from a large medieval manor house into Henry VIII's primary royal palace. His daughters, 'Bloody Mary' (r. 1553–58) and the future queen Elizabeth, were born there and it's where he signed Anne Boleyn's death warrant. When the time came, she was rowed from a wooden jetty in front of the palace upriver to the Tower of London, where she was executed.

By the end of the English Civil War of 1642–51, the palace was derelict. It was eventually demolished and replaced by Christopher Wren's Royal Hospital for Seamen, which admitted its first elderly and injured residents in 1705. It is these magnificent stone buildings that have watched over me as I have forensically searched the mud for evidence of the palace and its inhabitants.

As the sun falls and the temperature drops, the bright lights of the O2 Arena downriver illuminate a light mist that has settled over the water. It turns the beam of my head torch into a bright, solid staff that I wave like a lightsaber across the shingle and mud. It's much darker in Greenwich than it is in central London; there are no lights along the river path and people use mobile phones to light their way. I look up and see them bobbing past – the scene is timeless, they could be candles or lanterns. Gentle conversations drift down through the mist and an early drunk sings as he walks back from the pub.

I'm not scared of the dark; I'm more scared of being joined by a flesh-and-blood human. But here the dark plays tricks on me. Through the murk by the river wall come faces, grotesque and laughing. The wire nets of stones, set down several years ago to protect the base of the river wall, grin and gurn at me, and tiny baby faces giggle from the mud. Are they faces from the past? An old waterman with his hood pulled down low over his pock-marked brow; a young kitchen wench, throwing a bucket of bones and rotten vegetables into the water; a Spanish ambassador pulling up his collar against the cold, damp river mist.

It is a good night, and the river is beneficent. I find Tudor shoe soles almost straight away. Three of them are left-footed, one has a large hole worn all the way through, but only two are complete. I've often wondered if they were recycling leather at Greenwich, since I've found so many broken shoes and shoe soles but very few shoe buckles. Old shoes were valuable, and by the 1540s, they were even being collected for export to France, where they were sold into the second-hand trade and also refurbished and exported back to England.

There is a piece of a medieval jug handle and nothing else but pins. There are always pins at Greenwich, if you know where to look. Sometimes they wash up in thick tangles, other times they almost vanish altogether, but I can usually find at least one or two. A pin might be the most ordinary and common of all the objects I find on the foreshore, but even after all these years I can't resist picking them up. Each time I pick up a pin, I think of the effort it took to make it: drawing the wire to gauge and cutting it to length; wrapping another piece around the top three times to make a head that was sometimes soldered; then polishing and sharpening each one by hand. Before the pin machine took over in the early nineteenth century, countless pins were made this way. They have survived because they are copper alloy and not iron, which would have rusted away centuries ago. In the late sixteenth century, they were sold for about a penny a hundred, bought by the gross or thousand, and used by everyone from paupers to queens.

A woman's ruff or band was pinned to her rebato or pickadil; her rebato to the neck of her bodice; her partlet to the upper part of her kirtle, her kirtle bodice and skirt were pinned together; detachable sleeves were pinned to armseyes; farthingale pinned to petticoat; billament pinned to hood and coif pinned to hair. It took scores of pins each day to dress, and they were kept in pin pillows that were sometimes stuffed with emery powder to keep the pins clean and sharp.

By the eighteenth century, buckles and buttons had become cheaper, fashions had changed, and people used fewer pins. They bought them from poor and indigent street vendors who walked the streets of London singing 'Pretty Maids, Pretty Pins, Pretty Women', but according to the *London Chronicle* of 24–26 December 1761, English pins were still 'esteemed in all respects the best in the world'. Their reputation lay in the stiffness of the wire, but that was not without its risks. The way in which the wire was prepared meant that 'the slightest scratch by them will fester and prove very troublesome'. I'm always careful about picking up pins after reading that.

For centuries everyone prickled with pins and lost them in copious amounts. They can often be found near river stairs and causeways, where they fell from people's clothes as they climbed in and out of boats and wherries. If you know the locations of old drains, they can also be found there, where they washed into the river from streets and houses in the city. There is no way of knowing if the pins at Greenwich were once worn by monarchs or servants, or even if they came straight from the pin-maker's workshop. Archaeological digs have revealed evidence of a pinning industry in the area, and I have found several pinner's bones on the foreshore. These are cow or horse shin bones that were trimmed to size and had the knobbly bits at the joint end cut off to make them more comfortable in the hand and to sit flat on a workbench. The opposite end was squared off and thin grooves were filed into it to hold the pins in place while they were sharpened and polished. Pinners must have gone through piles of these specially adapted bones.

I only find three pins tonight and they are the last of my finds. I've seen out rush hour and with the river at my heels and the clock in one of the domed towers of the Naval College striking the hour, it is time to set off home.

Friday 18 March 2022
Booklarking

Today I am in Oxford to meet Francesca. She is the assistant librarian of rare books at the Bodleian Library ('the Bod'), which first opened to scholars in 1602. By virtue of her job – buying, cataloguing and curating antiquarian books – she is also a 'booklark', the finder of forgotten objects hidden inside books.

Booklarking has fascinated me since my mother gave me my great-grandmother's Bible. It is pocket-sized and almost as thick as a brick, with faded gilding around the edges and a tattered black-leather cover. The book itself is larked. The paper is probably made from recycled rags; the glue from boiled-up hooves and bones; the card, to which the thin leather cover is stuck, was quite possibly part of another old book; and it's likely the thin gilding was made from sweepings scavenged from the goldsmith's workshop floor. It is a cheap book made from borrowed parts and it sat on my bookshelves for several years before I paid it any attention. Then one day, in an idle moment, I picked it up and opened it.

On the first clean page, in fine curled script, my great-great-grandmother had dedicated the book to her daughter: 'Kitty Mail. A present from her affectionate mother, Christmas Day 1890.' Inside, hidden among the parables and psalms, I found five fragile leaf skeletons. They were just like the ones I used to collect with my mother when I was small, yet these leaves had fallen from a tree over 120 years ago and been saved by a woman I felt I knew but had never met. Had they been collected on a special day? Were they a gift from someone? My final finding on this tiny voyage of discovery was a 'secret' message written in pencil on the black end papers that was only legible by window light at the right angle. Someone, I presume my great-grandmother, had written 'My love is alive for thee. I will follow thee all the days of my life.' Most of what I had been told about Kate, also known as Kitty, was that she was an intensely private woman, not given to emotions or displays of affection, or even bouts of religious fervour, but these few words suggest otherwise. She had hidden a secret part of herself inside the book, and I had found it years later.

Since then, I've made a habit of flicking through second-hand books in junk shops and antique markets, which can feel like both a treasure hunt and a nose into someone else's business. I've found forgotten letters, postcards, photographs, tickets and newspaper clippings. Once I found an old-style £1 note. All fairly interesting but nothing to match Francesca's finds.

We meet in a bookshop opposite the Bodleian and settle down with a cup of tea. 'We find a lot of bookmarks,' she says. 'Sometimes they're specially made, often just scraps of paper, torn letters or scribbled notes. There are doodles and sketches made by bored readers. Even today students leave notes to each other in books, although it is discouraged.' Food for future booklarks, I think. I ask what kind of notes she's found, thinking about the old postcards and scraps of paper I'd discovered that enquired into health, told adventures of holidays and predicted arrival times for visits. 'We have an early seventeenth-century Bible, printed in Germany and bound with wooden boards. I noticed the endpapers were slightly loose and discovered the boards underneath had been hollowed out. Hidden inside was a folded note.' It was a speech made from the gallows by James Sheppard, who was hanged for treason around 1717 for 'imagining and compassing the death of His Sacred Majesty King George'. 'It's unlikely to be his actual written speech, though,' says Francesca. 'It's more likely to be a handwritten copy, hidden by a friend or supporter.'

I ask if the booklarked objects are kept in the library. 'There is a "janitor's list", a list of library objects that aren't books but are often related to books. There is a strict divide at the library between printed and manuscript, so anything handwritten found in a printed book has traditionally been separated and sent to the manuscript department and sadly the connection is lost.'

A lot has been lost over the years, especially old bindings, which were sometimes made from early medieval manuscripts that were bought in bulk by stationers and bookbinders to reuse as end leaves, to reinforce spines and line the inside of boards. The library has an enormous number of these cannibalised books, most of which haven't ever been looked at in any detail. Books were destroyed in the late sixteenth to eighteenth centuries too, when wealthy book collectors began to rebind their libraries in matching covers. Sometimes very old bindings were thrown away, including the metal clasps that held the old books closed. I have found quite a number of these on the foreshore and often wondered why. I'd been told that during the Reformation in the sixteenth century, outlawed prayer books and Bibles were thrown away, which offers another explanation.

I tell Francesca about the pressed leaves in my great-grandmother's Bible and ask if she's found anything similar. 'Pressed plants and flowers are quite common – we've found crocuses, violets, ivy leaves and oak – but the most common object is pins.' Pins had uses other than dressing and dressmaking. They could be bent into fishhooks and temporary clasps, but people mostly used them to pin notes into books, onto manuscripts, and to pin legal papers together. They were the precursors to paperclips and staples. The Bodleian's collection of booklarked pins dates from around 1600 to around 1899 and includes a pin that Jane Austen used on her manuscript of *The Watsons* (1805). They are kept pushed into small pieces of paper that are labelled with the shelf mark or collection where the pins were found and the date they were in use.

Before Francesca heads back to the library, I ask the same question that many people ask me. 'What's your best find?' 'Spectacles,' she says, 'probably early modern.' They were found inside a book of French love poetry, published in Paris c. 1526. She isn't sure if the Bodleian still has the actual glasses, but they have the book with the ghostly imprint of them on two adjacent pages and she says she can send me a photograph of it.

Before I catch my train, I have time to find the river. The Thames flows through Oxford as the Isis and there are several theories for how it came about this other name. One is that it comes from the ancient name for the river 'Thamesis', being Isis from its source in the Cotswolds until it is joined by the Thame at Dorchester in Oxfordshire. Another attributes it to students naming it after the Egyptian goddess Isis. In truth, nobody really knows, but I can't leave Oxford without visiting it.

I turn off the road just before the station and take a muddy track down to the canal. I don't have to walk far from the road for the sound of birds to cover the murmur of traffic. They are in full spring song, as bright and sharp as the sunlight. Greenfinches 'dweeeeeze', two robins battle each other with their sweet gurgling song, and great tits calling for their 'tea-cher, tea-cher, tea-cher' hop around in the bare branches of the trees. Bees, woken early by the warm spring sunshine, buzz among the bright yellow celandines that line the canal banks, and a yellow brimstone butterfly, always the first to wake up from hibernation, flutters past.

I negotiate a jumble of canals, streams and locks, and pass lines of brightly coloured canal barges with names like 'Golden Dancer', 'Land's End' and 'Promise'. There are bicycles and bags of coal on the roofs and on one an elderly, grizzled cat stretches lazily in the sunshine. I stop to look at glistening balls of frog spawn in the shallows and wish I'd brought a jam jar with me to take some home for the twins.

The river is in flood. It looks solid, a flat sheet dented by tiny eddies and whirlpools, level with the bank on my side and spilling into the flood meadows opposite. Geese, swans and ducks paddle noisily in the waterlogged grass, and I recognise the dark shapes of cormorants, surprisingly far from the sea, balanced in the branches of the crack willows along the bank. The river is smaller and younger than my London friend, but it has the same familiar mix of urgency and calm that draws me to it further east. It has around 115 miles to go until it reaches London Bridge, with weirs, locks and scores of bridges to flow under, villages and towns to pass through, and it will become a tidal river before it reaches the centre of London. I take a moment to breathe in its calm and to wish it well on its journey before I turn to retrace my steps back to the station.

Sunday 20 March 2022
Spring Equinox

Although the clocks don't go back until next weekend, today is the spring equinox, the official start of spring. Early people divided the year up astrologically between the equinoxes and solstices. In the days before calendars and clocks, it made perfect sense to live within these predictable boundaries of light, and it still does. Knowing we have reached the shortest, darkest day in December gives hope that light and sunshine are coming, and the equinox provides a tipping point into a new season. In London the equinox will occur at precisely 15.33, when the sun is exactly above the Equator, but the 'equilux', when day and night are of equal length, has already happened, on 17 March.

Each year, the spring equinox is a palpable point of relief, positive and life-affirming. I can feel the dark days at my back and sense a long bright road stretching ahead of me. From now until the summer solstice on 21 June, the days will grow and the nights will shrink. The change of time next weekend will gift another hour of light that will increase as the Earth tilts towards the sun and I will even have the luxury of searching two daylight tides in one day, should I want or be able to. Then it will slowly reverse, daylight will dwindle and the next tipping point will be one into darkness on 23 September with the autumn equinox.

Thursday 24 March 2022 (low tide 0.78 m @ London Bridge, 11.52)
Central London – North Bank

As I stand on the Millennium Bridge looking over the side rail, a dark cloud sweeps upriver and a curtain of rain falls from the sky, shattering the surface of the river and scattering the tourists around me. I pull up my hood and look down at the river.

I am assessing the lines of debris that are gradually being revealed on the foreshore below and working out where I am going to search. Something about the shape of the river and the currents means that the stretch between the Millennium Bridge and Queenhithe Dock on the north side is one of the best naturally curated parts of the foreshore.

The river is an obsessive sorter, and divides its booty up by size and weight. It scatters bones higher up, piles roof tiles together in humps and drops small metal objects in barely perceivable dips and hollows closer to the waterline. Knowing how to read the foreshore and understand how it sorts and shifts is a skill that comes with time and observation, but here the sorting is even more obsessive and obvious. Below me, there is a black line of coal, charred wood and cinders from ancient hearth fires, an orange fan of roof tiles and bricks and a buff smudge of bones and oyster shells. On every tide the river washes and sorts the clutter and adds something new, which is why, even though it's well searched by other mudlarks on most tides, it's usually worth having another look.

Foreshore bones are mostly animal bones from 2,000 years of London's dinners and waste from bones that were made into useful objects like knife handles and buttons. There are also the bones of animals that accidentally met their death in the river and those that were dumped in it when they had fulfilled their purpose. There are even the partly mineralised bones of prehistoric creatures, like red deer, elk, wolf and mammoth, but these are rare. Teeth also wash up with the bones. Sometimes they are blackened and worn, but usually they are still quite white and pearly, and I find them repellently fascinating. I have limited myself to collecting single teeth and only one example from each type of animal. I have pig, sheep, cow, horse, dog, cat and the front tooth of a rat, squirrel or rabbit, I'm not quite sure which. I keep the teeth in the same drawer of my collector's chest as the bone, horn, antler and ivory objects I have found on the foreshore. I think it is probably one of my favourite drawers. It's certainly the one I open the most just to take out the objects inside. Things that are made of bone are irresistibly tactile and beg to be touched.

In the drawer I have at least two handfuls of plain bone buttons of the type that were often used on underwear in the nineteenth century. There are game counters, dominoes, a small turned-bone pot missing its screw-on lid, ivory fan staves, beads, knife handles, toothbrush handles, shaving-brush handles, combs, a musical instrument tuning peg, dice, an eighteenth-century entrance token to a pleasure garden, two small spoons, an awl and a hammer made of deer antler, a lace bodkin, a parchment pricker and a decorated medieval ivory nail scraper. I have a lot of broken bone and ivory objects that I will probably never match with complete examples and even more bits of waste in the form of carefully cut sticks of bone, discs and tips of antlers, and flat pieces of bone with rows of circles and half-moons where buttons, beads and counters were cut out of them.

Close up, the drift is a jumble of gentle colours, soft browns, chalky whites, rich reds and black coal that sparkles in the sunshine. Bones that have been out of the mud for a time quickly dry to a soft brown once the tide retreats. Fresh from the mud, they are waterlogged, dark and sometimes even blue. The blue is vivianite, a hydrated iron-phosphate mineral that forms naturally on old wet bones and shells. I've even found it on old bone combs and knife handles. It is colourless until it oxidises, when it turns the deep, rich blue that was prized by some seventeenth- and eighteenth-century artists as a paint pigment.

I start at one end of the organised debris and walk slowly eastwards, bent as low to the ground as I can. I am looking for worked bone and also knuckle bones: the astragalus, an ankle bone, in this case specifically smaller ones from sheep, goat or calf. Apart from their satisfying size, shape and colour, I am convinced that I will one day find one with a pattern scratched onto it or the remains of paint that will confirm it was actually once used in a game.

The game of knuckles, fives or jacks has been played all over the world since ancient times with seeds, pebbles, bones and ceramic game pieces. In its simplest form, five knuckles are tossed up and caught on the back of the hand, but there are many different versions of varying complexity. I've found the colourful glazed ceramic cubes that Victorians used to play the game, but I like these simple small bones. I keep them in a glass jar on my bookshelf and when the jar is full, I'll stop collecting them; perhaps I'll take them all back to the foreshore. Maybe then I'll focus on finding a cramp bone. These are the kneecaps of sheep that were used as a charm against cramp. I first read about them in Dickens' *Bleak House*: 'Such mean little boys, when they were not dancing, with string, and marbles, and cramp-bones in their pockets...' Ever since, I have been looking for one, but thus far they have eluded me.

I clatter and crunch through shoulder blades, ribs, hooves, huge ox legs and fragile bird and rabbit bones. I pick up three knuckle bones and three circular spines from the skin of a thornback ray, but no cramp bones. A happy-looking mudlark nods as she walks past on her way to the river stairs and I wonder what is in her finds bag, and in the bags of all the figures I can see in the distance. Each one would be different. Different eyes notice different things; there are different interests and chance discoveries that can never be repeated. Is it even possible in this vast jumble to set out to look for something so specific as a cramp bone? Are we choosing the finds, or are the finds choosing us?

Friday 25 March 2022
The Pilgrim Badge

It was nearing dusk on a Friday evening a few years ago. I was on my hands and knees near London Bridge, on a completely deserted foreshore, when I noticed a small, flat, dark grey diamond-shaped piece of metal caught against a stone in a low water-filled dip in the shingle. A serene oval face below a bishop's mitre looked back at me and I knew instantly what it was. I had found broken medieval pilgrim badges before, but it had taken me almost two decades to find a complete one like this. My heart pounded, my mouth went dry, and I couldn't touch it. I knew that as soon as I did, my heart would cease to flutter with the magic of the find and my brain would take over, replacing wonder with facts: how old was it, who was the face looking back at me, how was it made, who lost it. I needed a few moments more with the magic before I scooped it up.

The first person I showed was Colin, who immediately identified the face as that of St Osmund of Salisbury. I found Colin online through his website, where he sells the beautiful reproduction pilgrim badges he makes. He was also recommended to me as an oracle on pilgrim badges by several medieval re-enactors. I consult Colin when I find little pieces of decorated pewter that I think might have once been pilgrim badges, and his knowledge really is encyclopaedic. He has recently identified the lower half of a fifteenth-century Our Lady badge for me, and a badge of John Schorne from just the pulpit. John Schorne was the rector of North Marston in Buckinghamshire, who is said to have saved his village by trapping the Devil in a boot.

The St Osmund badge was an unusual find for London. Although badges from as far away as France and Germany have been found on the foreshore, the most common are those of St Thomas Becket, whose shrine was in Canterbury Cathedral, a well-trodden four-day walk from London. My badge was bought by a pilgrim visiting a far less important shrine, much further away, at Salisbury Cathedral. They probably chose it from a pile of similar badges on a rough wooden table outside the cathedral and paid just a fraction of a penny for it. It was a cheap souvenir that may have been rubbed against the saint's shrine and worn on a hat or coat for protection, and to show others where the pilgrim had been. Perhaps they had bought one to wear and another to give to the Thames on their return. So many badges have been found in the river, it is thought they may have been ritual offerings, as thanks for a safe return and a vow completed.

Colin's collection of a hundred pilgrim badges includes around thirty from the Thames, which he says have a distinctive darker patina from the oxygen-free mud. But he doesn't just collect and identify medieval badges, he also makes them, and today I'm going to visit him to make a pilgrim badge myself.

Colin got into casting pewter badges about thirty years ago when he became a medieval re-enactor. He was casting models in pewter, but fellow re-enactors started asking for reproduction pilgrim badges to complete their outfits. He now has moulds for around 150 different religious and secular pewter badges, which he sells on his website. When he's not making badges or demonstrating how they are made at events, he spends his spare time researching and collecting originals. He mostly buys them from auctions and some from eBay, but he tells me you have to be careful because there are a lot of fakes.

I'd brought my meagre collection of bits to show him – St Osmund was in Maidstone on a shelf in Jo's corner – and I am itching to look through Colin's collection. He has laid out a selection for me to look at and there are boxes more of them, all neatly filed away in labelled plastic sleeves. 'One of the things I love about pilgrim badges are the stories behind them,' he says. 'Many of them are folklore, pagan stories turned Christian to inspire and frighten people into believing. Some of them are quite weird.' He picks up a large badge and holds it out to me. Five birds look as if they are sitting in a basket. 'They're geese and they're behind a fence,' he says. 'It's the badge of St Werburgh, an Anglo-Saxon princess turned abbess. A plague of geese had been eating her abbey's crops, so she instructed a servant to gather the geese in a pen so that she could deal with them in the morning. The next morning, she asked them to fly away, which they did, but they soon returned when they realised that one of them was missing. The night before a servant had cooked and eaten the goose, so Werburgh retrieved the bones from the kitchen, worked her magic and brought it back to life. Once they were reunited, the flock flew away and never returned.'

On another of Colin's badges, St Alban's head is hanging by its hair from a tree. 'There should have been an executioner next to the head, but he has broken off this badge,' says

Colin. Had the executioner been there, his eyes would have been missing, since he was struck blind at the moment of execution so that he would never witness the virtue of the martyrdom that arose from his actions. The badge of St Anthony may have been a protective charm against St Anthony's fire, a burning sensation in the extremities that eventually turned to gangrene. 'Today we know it's caused by eating mouldy rye grain, but back then they put their faith in a bit of pewter.' Then there was St Audrey, who died of a throat ailment, probably throat cancer, that she attributed to her love of jewellery and fine necklaces when she was a young princess. The souvenir necklaces that were sold at her shrine were so poorly made that they gave rise to the word 'tawdry'.

'Right, shall we have a go at making one, then?' he says, putting down his empty coffee mug and gesturing towards the back garden. On the lawn, a small iron cauldron is hanging by a chain from a tripod over a glowing charcoal brazier. The fire pops and crackles, and the deliciously evocative smell of smoke rises into the air, transporting me back in time to the workshop of a medieval pilgrim badge-maker.

The pot is filled with a silvery liquid metal, reflecting brightly in the sharp spring sunshine. It is mesmerising and I want to touch it. 'This is a traditional mix of tin and lead, just as they were made centuries ago, but the ones I sell are safe pewter without any lead.' Colin taught himself how to make the moulds, and a selection of finely carved limestone examples are lined up on a wooden bench beside the brazier. English pilgrim badges usually had pins on the back and required three-part moulds, while continental badges had loops to sew them onto clothing and could be made in two-part moulds. Small pegs and corresponding dips help to locate the pieces together and there is a wide channel at the top, a pouring canal, where the liquid pewter is poured in. Minute channels spider out from the carved design to the outside of the mould. These are tiny chimneys, vents as fine as a hair, that allow air to escape as the metal finds its way in and around the intricate designs.

Colin picks up a large St Thomas Becket mould, featuring Becket seated and holding a crozier. He dusts it with finely powdered stone to stop the metal from sticking, pulls on a pair of thick leather gauntlets and, holding the mould in one hand, dips a well-used ladle into the shining liquid. He deftly pours just the right amount into the hole at the top and flicks his wrist to knock off a tiny bit of excess. In seconds, the cold stone has absorbed the heat of the molten metal and it is set. He opens the mould, knocks out the badge and breaks off the casting sprue with a quick wiggle. In that moment another foreshore mystery is solved.

Years ago, I found what looked like a little lead tree. It wasn't very exciting, and I only kept it because it looked like it should be something. I'd get it out sometimes and turn it over in my hands, sit it on my desk beside my computer for a few days, but I never worked out what it was. Then, as Colin knocks another badge out of the mould, I finally realise what it is... a casting sprue, waste pewter that had hardened in the pouring canal of a mould. 'They are probably rarer than badges,' says Colin. 'They would have been thrown back into the pot and melted back down again.'

He hands me a freshly made badge. It is still warm in my hand, and I can clearly see Becket's large head balanced on his feeble shoulders and his trademark downturned mouth. This is just how my St Osmund would have looked when it was bought, I think. 'They dull down quite quickly,' says Colin, 'but don't they look great new?' In a world of greys, browns and greens, I can see why people bought them. This shiny little thing was a piece of the magic, wealth and awe of the church.

He takes the badge back, bends the pin catch into a hook and runs a knife around it to clean up the ragged edge where the metal has flowed into the tiny vents. 'They didn't always bother doing this – cheap and cheerful they were.' It is perfect, so perfect in fact that Colin casts the word 'REPLICA' into the back of his badges 'To stop people from burying them for a few years and passing them off as real.' Then it is my turn. I choose a mould for a smaller Becket badge, just his head and shoulders in a circular frame, and repeat the steps I'd been shown, but far less expertly. It works though, and I tap out a perfect badge. I make a couple more and take several home with me. Like a medieval pilgrim I pin the small Becket badge to my mudlarking jacket. Perhaps it will bring me luck, maybe it will protect me; really, I just want to see how long it takes for it to lose its shine.

Tuesday 29 March 2022 (low tide 1.15 m @ London Bridge, 19.09)
Central London – North Bank

My first summer-hours lark is cold and damp, and I don't stay long. The low tide isn't good and the best searching spots stay under water. I am forced to search parts of the foreshore near the top of the wall that I don't usually bother with, but that turns out to be a good thing, maybe fate. I'm a great believer in fate. Fate has led me to be in the right place at the right time to find objects that have surfaced after centuries underground, which means every single thing in my collection is meant to be there: they are all fated.

Today, I find a cast, conical lead weight, with a hole drilled in the top, which is probably an old fishing weight, possibly post-medieval, and a very worn and bent copper coin, smaller than a modern penny, that turned out to be a French Henri II (r. 1547–59) *denier tournois*, the oldest foreign coin I've ever found. Then I drop onto my knees just before Southwark Bridge, where people have already churned up the black gritty mud with their boots. Where the tide is pulling back over fresh foreshore and the sand is smooth, I spot a near-perfect grey circle and lunge forwards to snatch it up. It is little more than a fish scale, a tiny Charles I halfpenny with a Tudor Rose on each side that had slipped from a purse or wet, frozen fingers and fluttered down into the mud some 400 years ago.

Such minuscule coins must have been very easy to lose. Perhaps it was dropped by a passenger paying a waterman, or by the waterman himself. The Thames would have been swarming with little boats at the time, wherries ferrying people up and down and across the river. An Admiralty muster, or census, of 2 February 1628–9 recorded 2,453 watermen belonging to 'the Port of London and the liberties thereof' (from Windsor to Gravesend), but half a penny wouldn't have got you very far. The first time the watermen's fares were fixed was in 1514 by an Act of

Parliament. Another Act in 1555 empowered the Lord Mayor and Aldermen of London to set fares themselves. In 1559 they drew up a table of fares that remained unchanged until 1673, when fares tripled. At around the time my little coin was lost, the fare for crossing the river from the north side, where I found it, to the theatres, brothels and bear-baiting rings on Bankside opposite was 1d (one penny) in your own boat, or ½d as a passenger in a shared boat. The return journey from London to Greenwich in a wherry was 8d rowing with the tide, and 12d against it.

As I jostle through the crowds on London Bridge on my way back to the station, I pause halfway and lean out over the bridge wall. Below me a tugboat slides upstream with the tide. The water is flat and brown under the cold white sky and when I pull one of the shiny pilgrim badges I'd made with Colin out of my pocket it blinks and glitters in the weak sunlight. St Osmund had been with me all day and perhaps he'd had a hand in sending me further up the foreshore than I intended, to find objects so minuscule they would otherwise have remained undiscovered. I stand on tiptoe and think about all the unfound things in the river and how fate has yet to unite them with their finders, then I close my eyes and, like a returning medieval pilgrim, I drop my shiny badge into the river and give thanks.

CODE: 124.22.GW07

OBJECT:	12th century penny, cut in half
MATERIAL:	Silver
DATE FOUND:	12/04/2022
LOCATION:	Central London – south bank
NOTES:	A medieval silver short cross penny, cut precisely in half along the line of the cross to create a halfpenny. Minted in Norwich or Northampton in the reign of King John (r.1199–1216). Surface find.

APRIL

Cornwall

'Look out for the fairies!' I shout over my shoulder to the back seat, but my enthusiasm for the little people is met with a stony silence. The twins are almost ten and I feel like I am losing them. I shouldn't say losing them: I am losing the little versions of them that believed goblins rode hedgehogs at the bottom of our garden and screamed when they heard tree branches creaking in the wood because 'the Old Man of the Woods was coming'. They used to believe, if they looked hard enough through a hole in a hag stone, they would see mermaids out at sea, and when I told them to look for fairies on the steep mossy banks that lined the impossibly narrow Cornish roads that led to my aunt and uncle's farm, they would press their noses to the back windows and look as hard as their little eyes could look.

I wind down the window for some fresh air, and the familiar smell of wild garlic and primroses fills the car. Beside me, Sarah is staring straight ahead, a muscle in her jaw is twitching and she is silently gripping the seat. Even after sixteen years on this side of the Atlantic, these narrow country roads still make her nervous. I grew up with them and I am not too worried about meeting someone else coming in the opposite direction. The lane is so rarely used, there is a strip of grass and herbs growing in the mud down the middle of it.

We nudge our way carefully through the narrow, ancient, wooded valley that has been cut into the landscape by generations of hooves, boots and cartwheels travelling from village to village, farm to field to market, and on Sundays to church or chapel. Some of these ancient tracks, or 'hollow ways', are so deep they have their own microclimate and were so busy that in the past farmers put straw on the roads to take up the animal dung and used it on neighbouring fields. It wasn't until quite recently, within the last hundred years, that their sharp descent into the granite-flecked loamy red soil was halted by a layer of tarmac.

The farm is in a lush Cornish valley that's the next best thing to heaven. There are barn owls in the old farm buildings, bats that flit across the lawn at dusk and newts at the bottom of an old stone trough by the house that can be scooped out with nets and admired in jam jars. The house is made from granite, dug from a quarry in one of the fields, and is so ancient it's mentioned in the Domesday Book. It was a working dairy farm until my cousin Robbie, who was next to take it on, died suddenly and the cows were sold. The farm is sad and quiet now, the parlour is closed and dusty, the barns are filled with junk and there's only the faintest familiar smell of cows when it rains.

When I was young, everyone in my father's family lived on a farm and they always had, for hundreds of years. Not big farms, just family farms with the same scrap heaps, silage pits, smells, tractors, dogs, mud and wellies at the door as ours. The land connected us all, way back through generations. There are old photographs of my ancestors harvesting, holding prize chickens and standing next to hay ricks, their hands on working horses and cattle, squinting into the sun from the seats of old-fashioned tractors. My cousin says there are more photographs of cattle in our family album than there are of people, and she's right.

I have my great-grandmother's silver cup for 'Best Couple of Cockerels, Uncrammed any Breed' that she won at the Dairy Show in 1928, 1931 and 1932. I also have an antique oak chest or 'coffer' that's been passed down to daughters and 'held' or 'lent' to sons in lieu of a daughter since at least the early 1700s. The wide planks are held together by wooden pegs and the oak has aged to a rich dark brown from years of beeswax polish and candle soot. Its legs are pocked with ancient woodworm holes and it is simply decorated with stylised leaves that wind down the sides and legs. I pulled it away from the wall once to look at the back, which is plain with rough saw and chisel marks from the joiner's tools. It is a very ordinary chest, of the type that most people had in their homes to store their meagre posessions and I often think about what my people kept in it over the years: candles, embroidered linen, quills, pewter plates, winter coats, Bibles, hard-earned savings, important documents, shoes that were only worn on special occasions, precious possessions that are all worn out, lost and scattered to the winds now, while the most precious possession of all, the oak chest, has stayed with us.

It came with a piece of paper that listed some of its owners, not just by name but by farm. Grace Yard from Currey Mallet in Somerset owned it in the early 1700s; she passed it to her daughter Keziah of Radigan Farm, then it was held by a son, Walter, for his daughter Florence of

Spalsbury Farm who lent it to her brother Joe at Bridgefoot Farm, who gave it to his daughter Daisy (my grandmother) at Guiles Hill Farm, from where it was lent to my parents at Dean Farm, for me. In time I will pass it on to my daughter, but I doubt it will ever go to live on another farm.

We have always been a farming family, and when I was young I assumed we always would be, but in one generation (my generation) the farms have been sold or given up and now it is my aunt and uncle's turn. They are moving to the village nearby and this will be our last visit to their farm. We arrive mid-chaos. A lifetime is being sorted out and packed up. There is tension in the air, and we don't want to get in the way, so we say hello, drop off our bags and take ourselves off on a familiar walk, up Joe's Lane to the village to buy ice creams.

The walk takes us past Joe's Farm, Joe's barn and Joe's dump. 'I'm just going in here,' I say, as I dive into the bushes. I found the old dump a few holidays ago. My cousin Julian told me it was probably another old quarry, for the stone that built Joe's farmhouse nearby. Presumably Joe had also added to the stuff that could be found in the undergrowth. Before the advent of organised rubbish collection, most houses had a dump or midden where household waste was disposed of. It may have been a pit at the bottom of the garden, a dried-up pond, a ditch or an old quarry like this, anywhere it was easy to get rid of what they didn't want.

The contents of this dump fascinate me. It is a glimpse into a stranger's life that makes me feel excited, curious and nosey all at once; there's also the chance there might be treasure in the undergrowth. The rubbish all around me probably dates no further back than the late 1800s, but judging from the plastic bottles and containers that lie around in the undergrowth, it was still being used well into the 1980s. Most of the glass bottles and jars have been smashed, but I recognise a mid-twentieth-century medicine bottle, with doses marked out in lines in the pressed glass. A hart's-tongue fern is growing inside an old-fashioned jam jar with a lip for tying a square of cloth or paper over the top, and there are several squat brown Bovril beef-extract bottles. There are always Bovril bottles and Shippam's fish-paste jars in old dumps.

The twins climb up the bank and join me. They poke around for a while with sticks then Twin 1 jubilantly holds up a dirty white enamel kettle, rusted through with holes. Inside, it is filled with large buff-coloured hibernating garden snails, and a cache of empty hazelnut shells tumbles out. Each one has a delicately nibbled hole in it, so the kettle had been a shelter for snails as well as a wood mouse's larder. We carefully nestle the kettle back in the dry leaves where we found it, then Twin 2 calls across to us, 'Look at this!' This is our treasure. It is an old leather boot, probably Joe's boot, tailored with soft green moss that looks as if it had been woven into the leather itself. Nature is claiming the dump, piece by piece, making use of what had been thought of as useless, and turning ugly things beautiful.

We crouch down together to look closer at the boot. There is a thin line in the moss that follows a line of stitching across the heel, and it has grown around the empty lace eyelets. 'You know what?' I say. 'If the Old Man of the Woods wore boots…?' and two little faces look back at me with eyes as wide as saucers.

Saturday 9 April 2022
Shared Memories

Among the pile of bills and circulars lying on the doormat when we get home from Cornwall is a Jiffy bag forwarded from my agent. Inside, in a small box and carefully wrapped in tissue paper, is a fairly unremarkable piece of flint. The accompanying letter is from a woman in Quebec who explained that she found the flint on the banks of the St Lawrence River near where she lives. She went on to say that there is a lot more flint to be found on the river and that it is thought to have arrived as ballast in ships that came from the Thames. So this ordinary stone isn't so unremarkable after all. It had travelled thousands of miles by sea several centuries ago and taken a relatively short plane ride back home. I put the flint on the telephone table by the front door to remind me to take it back to the river where it belongs.

Since I started posting my finds online, people have been sending me their memories. It seems you can leave the river, but the river never really leaves you. People send me their stories of playing on the foreshore as children, going down to the river with now departed grandparents and parents, and special days when they saw magical things and found wonderful objects. Their stories are like finds, they wash in on my computer or in the post, and I snatch them up before they're gone.

People also show me photographs of objects that they've found, the most exquisite of which was a silver brooch. It was in the shape of a swan with a large Baroque pearl for its body in a style I recognised as probably sixteenth century. The sender's father had found it on the foreshore during World War II and its discovery had pre-dated the Treasure Act that requires such objects to be reported by law. It hadn't been recorded with any of the museums and very few people had seen it, so in essence I suppose it doesn't really exist. She said it was a family piece now and that they didn't want to risk it being taken away. I understand, but I hope it is passed on through the family along with the story of its discovery and where it came from.

In 2020 another Jiffy bag arrived. Inside was a letter and a package labelled 'FRAGILE MUDLARK FIND 1979'. The author of the letter apologised for writing out of the blue and introduced herself as an eighty-year-old woman who had read my book and felt inspired to share the memories it brought back. She had lived for much of her life beside or close to the river, growing up in Hammersmith and mudlarking there with a schoolfriend in the 1950s and 1960s. In her mid-teens her parents moved to Rotherhithe to run a pub and she continued her forays on the foreshore, making her way through docks and jetties to reach the river. She married her 'mudlarking soulmate' and they lived in Vauxhall for seventeen years, where they searched the riverside together for treasures.

I opened the package she had sent. Inside was a ceramic Victorian Cherry Toothpaste pot lid from the Army & Navy Co-operative Society, a member-owned department store that was founded in 1871 by a group of military officers. It was broken in half and had been carefully glued back together. Her letter explained that she had found one half on the foreshore while her husband had found the other half over a mile further upriver. Written on the back of the lid in black ink was: 'THIS WAS FOUND IN TWO PIECES, APPROX. 2 WEEKS APART IN THE RIVER THAMES AT VAUXHALL. 1979.'

She wrote that her husband had died and that she was too old and frail to mudlark any more. Since she didn't think her family would be interested, she had decided to send her 'impossible trophy' to me, to return to the river or to keep, whatever I saw fit. I wrote and thanked her and told her I wasn't going to throw it back in the river. How could I? I've kept the impossible trophy in my finds chest, along with the letter and a weight of responsibility for someone else's story.

Tuesday 12 April 2022 (low tide 1.96 m @ London Bridge, 17.14)
Central London – South Bank

It feels good to be going back on the foreshore and comforting to run through my methodical pre-larking preparations. I get out my filthy rucksack, check the tides and train times, make sure I have latex gloves, knee pads, finds bag, water and a snack, then I lay out my clothes and boots ready for a quick getaway. Although I miss it, it's sometimes good to have a break from the river because I appreciate it more when I come back to it: smells are stronger, colours are richer and the sounds are fresh and new.

The foreshore is sunny and busy when I arrive, but once the wind gets up and blows in some heavy showers, the random interlopers and curious tourists soon flee for shelter. Just a French couple braves the rain and asks me where the next set of stairs are. I watch their brightly coloured waterproofs vanish into the deluge. A noisy mallard duck paddles past in the shallows, followed by nine delicate balls of fluff that rock and struggle in the waves behind her. They are newly hatched, and I wonder where on this heaving tidal beast their mother had found a safe place to nest. Then I turn my attention to the foreshore. Since I have been away for a while, I need to get my eye back in. 'Getting your eye in' is what mudlarks call the honed ability to pick something out of nothing based on its regularity. Mother Nature doesn't make many perfectly straight lines or circles, and they stand out among her natural irregularity. In this way, a trained eye will spot a coin among pebbles or a pin in the sand.

At first, my eye skitters over the foreshore; I am looking at too much and moving too quickly. I need to slow down and focus. I work my way west along Bankside, gradually slowing my pace until by the time I reach a good area of erosion, I am relaxed enough to spot a thin white sausage of clay sticking out of the gritty mix of ancient road sweepings, household waste and hearth ashes.

It is a clay wig-curler that had once been dumbbell-shaped but has broken in the middle at its narrowest point. Most of the wig-curlers I've found are broken; only two are complete, which suggests that they probably ended up in the river with the rubbish. Clay wig-curlers were used in the seventeenth and eighteenth centuries to set hair on the large wigs that were fashionable at the time. Cheaper wigs were made of horse or even yak hair, but the finest wigs were made of human hair. A man could outfit himself with a hat, coat, breeches, shirt, hose and shoes for the price of a good wig, so they were well looked after and there was quite a skill to making them. In his book *Plocacosmos, or the Whole Art of Hair Dressing* (1782), hairdresser James Stewart explains how the curlers were used and the wig was baked in a pie to set the curls:

> *Hair which does not curl or buckle naturally is brought to it by art, by first boiling and then baking it in the following manner; after having picked and sorted the hair, they roll them up and tie them tight down upon little cylindrical instruments, either of wood or earthenware... in which state they are put into a pot over the fire, there to boil for full three hours; when taken out they let them dry, and when dried they spread them on a sheet of brown paper, cover them with another, and then send them to the pastry-cook, who making a crust or coffin around them, of common paste, sets them in an oven till the crust is about three fourths baked.*

After the wig-curler, the river continues welcoming me back with a wealth of finds: buttons, beads, pins, some beautiful shards of pottery, a broken medieval floor tile and a poorly cast counterfeit George III farthing. By the river wall, in a patch of nails and iron, I find a silver coin, neatly cut in half. I recognise it as medieval from the cross on the back that is short and not long. This means it had been minted before 1247, when the length of the cross was extended to the outside of the coin to stop people from stealing silver and devaluing the coin by clipping around the edge. If the ends of the cross were clipped, the coin would no longer be worth a penny.

At the time, coins were worth the value of the weight of the metal they were made from, and the lowest denomination was a silver penny. To create small change, pennies were cut in half to make a halfpenny and into quarters for farthings. This coin is so neatly cut in two it can only have been a homemade halfpenny, but I have no idea who the king is. He would have been hard enough to recognise even with a complete portrait. Like Roman emperors, all the medieval kings look the same to me. All there is, is the top of a head with some curly hair and an indistinct crown, which makes it impossible, or so I think, to work out who it is.

Back at home I email Paul. He is the man I contact when I find a coin that's difficult to identify and he never fails to amaze me. He can identify the medieval kings even when the coin is worn or broken. And not just that, he can also reel off the Latin legend around the edge and identify the mint mark that shows where the coin was made. Paul replies with his process of elimination: 'The main clue is in the words on the reverse', he writes. 'I can read ON:NORV, which translates as "at Norwich", or it might be ON:NORH, which would attribute it to Northampton. The style of the lettering takes me away from Henry II (r. 1154–89), Richard I's (r. 1189–99) coins almost always show a crown with seven pearls, yours has five, and Henry III (r. 1216–72) did not issue Short Cross pennies from Norwich or Northampton. This leads me to suggest, with reasonable confidence, that it is most likely King John (r. 1199–1216).'

King John was nicknamed John Lackland for losing the Angevin-Plantagenet lands in France. He so crippled

England financially that the barons rebelled against him and forced him to sign the Magna Carta in 1215, on the banks of the Thames at Runnymede. More popularly though, he is evil King John in the tales of Robin Hood.

The Magna Carta, or 'Great Charter', limited the king's power and prevented arbitrary royal acts like land confiscation and unreasonable taxes. It is around 3,500 words long, took fifty hours to write and there are sixty-three clauses, two of which refer to rivers. Clause 33 ensured the removal of fish-weirs from 'the Thames, the Medway, and throughout the whole of England', and clause 47 called for enclosed riverbanks to be made ordinary land again for the people to use. Fodder indeed for fighting those who would claim the Thames foreshore as private land.

I sit at my desk and hold the tiny half-moon in my hand. What would have been the effect of such a loss on an ordinary working man or woman? In the year 1212 the maximum daily rates for those employed building Salisbury Cathedral were:

Carpenters: 3d per day including their food,
4d without food.
Assistants: 1½d per day including their food.
Masons: 4d per day (master mason 6d a day).
Labourers: 1½d per day.

With their wages they could have bought 4½ lbs (2 kg) of wheat or a bottle of wine for 1d, a duck for 2d or a chicken for 3d, depending on size. Unlike the rest of the objects I took home today, this wasn't unwanted waste or something someone wanted to get rid of, like the counterfeit farthing, it was a significant random loss, a third of a labourer's daily wage, or a meal for a family.

Monday 18 April 2022 (low tide 0.45 m @ London Bridge, 10.14)
Rotherhithe Peninsula

According to the Elizabethan antiquarian John Stow, in 1114 the tide ebbed so far 'that between the Tower of London and the bridge and under the bridge, not only with horses, but also a great number of men, women and children did wade over on foot'. There was another magnificently low tide in the nineteenth century, during a very hot summer, when the tide fell abnormally low and lines of blackened 'tree stumps' were seen where the Roman bridge is thought to have been. Tide control and rising sea levels means we are unlikely to ever see tides like these again, but there are still days when the tide is low enough to attempt routes that mostly remain impassable. Today is one of those days, so I decide to walk around the Rotherhithe Peninsula, down its western leg to the Mayflower pub.

I'd visited Rotherhithe soon after the first lockdown was lifted in 2021. It was a sweet return after the longest stretch of time I had been away from the river in years. During Covid, the Port of London Authority closed the river to all but anything essential and the Thames was quieter than it had been for over 2,000 years. Only the occasional tug, PLA craft or police boat slid past on the smooth, abandoned water and mudlarks were told to stay away.

I relied on other people's eyes to show me what was happening. Photographs taken on daily sanctioned walks from bridges showed how clear the water in the shallows had become, and thick sludge was building up on parts of the foreshore that had otherwise been washed clean by boats, covering any finds that might have been there. People said the river seemed slower, that it had settled, and I imagined a thick, viscous tide, heaving up and down from the tidal head to the sea. The world had paused, the urgency had gone from the river and for a moment time stood still.

I heard that in the absence of the city's usual smells of restaurant cooking and pollution, they could smell the river rising up in dusty wafts to the bridges. In my mind I saw scented tendrils fingering their way along empty streets and down abandoned alleyways. Without the traffic and usual hubbub of the city, I wondered if the sound of the river travelled too, if the gentle lapping of the tide on the foreshore could be heard at the door of St Paul's Cathedral.

It was agony knowing that the tides were rising and falling without me. I wished I could be there to see it for myself and watched with envy and awe as the mud and silt on the foreshore turned acid green with algae, and nature began to return. House martins were seen nesting under concrete jetties. Their cup-shaped nests were fashioned from the algae and from river mud that contained minute fragments of London itself: brick, ceramic, flakes of iron and fragments of bone. Seals were regularly spotted as far up as Richmond and wading birds usually only seen on mudflats much further east moved upriver towards central London. Water birds of all types were enjoying the peace and calm; even herons were now stalking the clear shallows at Wapping.

I chose Rotherhithe for that first post-lockdown walk because I could get to it without using public transport and I also thought it would be quiet, but I was wrong.

While the river was calm and flat and the sky was empty of planes, the foreshore was heaving with people, more than I'd ever seen before. Swimsuits and flip-flops were on, shirts were off, and coolers were out. Thin wisps of smoke rose from disposable barbecues and a group of children were building sandcastles on a patch of sand. A sunburnt man was sitting in a folding chair between two long fishing rods, and next to him three Canada geese were preening themselves at the water's edge. A duck glided past, eyeing the scene warily, as a group of boys cavorted in the shallows, stripped to their pants, and a woman paddled by in a bright white bikini. More noise came from the apartments in the converted warehouses above and loud party music was pumping out of windows all along the riverside. The sounds mingled in the air and fell onto the foreshore, where it joined beach noises, the hubbub of conversation and squealing children.

As I picked my way between the bodies, I began to feel decidedly overdressed in wellies, rubber gloves and a backpack, and I realised people were staring at me. I wasn't surprised. In all of this weirdness, I was the weird one. I was the odd one out in the place where I had always felt so comfortable and at home. I tried to focus on what I was doing, but as I bent down to pick up an eighteenth-century clay pipe bowl, I overheard two people on a balcony above, loudly discussing exactly how weird they thought I was.

'What's that person doing? It's weird,' said one. 'I don't know, they're here at five in the morning sometimes,' came the reply. 'Weird.' 'Yeah, fucking weird,' came the consensus.

But today I have Rotherhithe to myself. I begin my walk just east of the tip of the peninsula and turn down a narrow, paved path that leads to a stretch of the river known to mudlarks as 'The Farm', because it's in front of Surrey Quays City Farm. On the other side of the tall, ivy-covered wall the animals are waking up and I can hear their sleepy, early-morning grunts and straw-shuffling.

I am early and the first person on the foreshore. There are no footsteps in the mud, so I know it is virgin territory. The tide still has a way to go, but it has dropped low enough to start searching among the ships' nails and large 'U'-shaped iron staples that litter the foreshore at South Wharf. The staples were used by men called 'rafters', along with ropes and metal straps, to lash cargoes of wood together that were then stored secured to moorings, floating on the river.

As the tide drops, it reveals old car tyres, broken beer bottles and ships' timbers. Heavy chains and iron cables emerge from the shingle and thick ropes slither from the mud like giant snakes, curling and turning before they disappear back into their sludgy lair. There is a lot of rubble, a broken white ceramic butler's sink, a slab of white marble that looks like the top of a table or counter, smashed Victorian-stoneware lemonade or ginger-beer bottles, and a shard of glass from a Pepsi bottle with an old-fashioned 1960s red-and-white logo on it. I catch a whiff of oil and tar as I bend down to pick up a clay pipe. When I wash off the mud in a nearby puddle, I see it has a maker's name stamped into the bowl, 'FORD STEPNEY', with the symbol of a bee underneath the name.

The pipe dates from 1820–40 when South Wharf was a depot for wood. Maybe it fell out of the pocket of a rafter. It was made by the Ford family, who had a pipe-making works at 49 White Horse Street in Stepney. The Fords' workshop was just a two-minute walk from where my great-great-great-grandmother (on my mother's side) Eliza, a dressmaker, was living with her husband John, who was a dockworker. I wonder if they knew the Fords and how many times a week they passed by the workshop. Did my great-great-great-grandfather smoke a pipe with the figure of a little bee pressed into the bowl?

The pipe is lying among the mangled remains of the South Wharf Receiving Station, where people with contagious diseases, such as smallpox, scarlet fever, typhoid and diphtheria, were taken to await transfer to an isolation hospital further downriver at Long Reach in the Kent marshes. The receiving station stood beside the river from 1883 to 1940, when it was bombed and set ablaze during the Blitz and its remains were scattered onto the foreshore. I've always wanted to find a link to this lost hospital, and lying among the bricks and rubble I see it: a rather nondescript shard of white pottery with LCC (London County Council) on the bottom. It was probably the base of a clunky utilitarian bowl or cup that served tea, soup or simple meals to patients as they waited for the ambulance steamer to take them away.

I zip the shard and the pipe into my finds bag and turn to assess the tide. I have to time this walk perfectly to avoid getting cut off, and it looks like a good time to start walking. The mud can be deep in places on the peninsula, so I feel my way along slowly, testing it before I commit and using piles of rubble and ships' timbers for stability where I can.

I splatch on through the mud, passing a huge steel wall set into the riverside. The river has stolen the paint from the bottom, but the top is still bright yellow with bold red letters that read:

ENGINEERS
MILLS & KNIGHT LTD
NELSON DRY DOCK
SHIP REPAIRS

It is hard going, but I soon reach the tip of the peninsula, the nub of the elbow that is a remote expanse of stable yellow sand where I can rest. I look out across the river to the towers of Canary Wharf as a crew of eight rows past against the tide. I am blissfully alone and a world away from the city and the glass and steel opposite me. But it is not a pristine beach. A bright orange sari is wrapped around an old wooden post and a triangular 'men at work' road sign is stuck to the surface of the sand. A sand-filled jacket, like the prone torso of a man, stops me in my tracks, then I notice a dark hump in the distance. I squint at it and see that it is looking back at me. The seal doesn't like me being in her space and as I back up slowly, she heaves herself into the water and slips away, leaving me completely alone again.

After a while I carry on walking west and soon spot someone kneeling on the foreshore in the distance, the only other person I've seen all day. As I get closer I see a familiar face, a mudlark I know called Tom. He tells me he should really be working but has been lured to the foreshore instead.

I tell him I am walking all the way around the peninsula and he wishes me luck: 'I've thought about doing that, but I've never felt brave enough.' He plants a seed of doubt, and I wonder if I should carry on. I have already navigated a couple of tricky spots, but I still have to get past the old Surrey Basin entrance, where the mud can be very deep. I look at my watch and over at the river where the tide is still going out. 'I think I'll be all right,' I say. I can always double back if I have to.

I say goodbye to Tom and carry on, stooping under jetties, sliding over rubble and shuffling in tiny steps along slippery concrete revetments. When I reach the Surrey entrance, I take it at speed and make it across the mud without getting stuck. According to my phone I have walked seven miles to my destination and I am standing on the stretch of foreshore that had been so busy nine months earlier. This time there are no crowds, no barbecues or sunbathers, I am the only person here and the foreshore is quiet, as it should be.

Wednesday 26 April 2022 (low tide 1.69 m @ London Bridge, 17.56)
Central London – South Bank

I have a meeting in town today and manage to sneak down to the river for a quick look afterwards. I don't have much time before the tide covers the foreshore, but it is a good place to unravel and digest after a long meeting. I pick up a post-medieval bone knife handle, a farthing from 1917 and a shilling from 1953, the year high tides and a storm surge in the North Sea combined to inundate the estuary, flood homes and drown people and animals. I also find another leather inner sole. It is black from being in the mud for so long and the leather shines softly from where a stockinged foot slid against it with every step it took through London's streets. The irregular holes around the edge tell me it had been hand-stitched, and I estimate it to be a good couple of hundred years old. It is also a left shoe.

I often wonder how all the shoes I find ended up in the river. Were they lost or had they been dumped? Even today countless lost shoes languish on railway tracks, beside roads, in parks and on beaches. I've read that it is more often left shoes that turn up on their own. It is a phenomenon, but why? Is it because most people are right-footed, and the left shoe gets left behind?

I look through my collection of river-found soles when I get home. I keep them in an old shoebox after I've dried them, pressed like flowers between newspapers, weighted flat with a pile of bricks under a sheet of hardboard. Of the sixty-four soles I have, twenty-five are right-footed and thirty-nine are left. In the pile there are Tudor soles shaped like wide-toed duck feet; the soles from delicate, narrow-waisted medieval ladies' shoes; thick Victorian boot soles with hobnails; thin, stylish Edwardian soles; a pointed winkle-picker from the 1950s; and even a ballet-shoe sole that I found years ago, way out in the middle of nowhere in the estuary. There are soles from 'straights', shoes that were common between 1500 and 1700 and not cut specifically for the right or left foot. Some of the soles are very large, others are from tiny baby shoes. Many of them are worn right through and these are the ones I like best. Shoes keep memories in their wrinkles and creases, and they all have stories to tell.

CODE: 55.22.BB05

OBJECT:	16th century posey ring
MATERIAL:	Silver
DATE FOUND:	05/05/2022
LOCATION:	Central London – south bank
NOTES:	A silver finger ring decorated with a repeating pattern of lozenges, lines and dots that may have been enamelled. '+ALWAYES' engraved on the inside in block capital letters. The ring has been cut and squashed, perhaps in an attempt to remove it from a finger. Reported as Treasure. Surface find.

MAY

I don't sleep well, and I'm often up with the lark. In the dark winter mornings I can sometimes get back to sleep, but as they get lighter, sleeping gets harder. I've largely given in to my summer insomnia and learned to love daybreak when the world is still and silent. This morning I make myself a cup of tea and sit outside wrapped up in a blanket, while the birds sing me a May Day chorus. There are no larks here, but I can hear swifts screaming and look up to see the black sky-racers wheeling and swooping as they chase their insect food. Their name comes from the old English *swifam*, meaning 'moving fast', and they are high today. This means good weather: the higher the swifts, the higher the pressure and the finer the day.

Swifts are strange, otherworldly birds that are rarely seen up close and spend their lives as mysterious black dots in the sky, sleeping and mating on the wing, and catching raindrops to drink. The ones above me are early, back from the savannahs and forests of Africa, a journey on which their uselessly short legs barely touch the ground. Swifts have little need for legs. Even their Latin name reflects their skyborne nature, *Apus apus* (*a* – 'without', *pus* – 'foot'), since people once believed they had no feet at all. They also thought they vanished into mud or pools for the winter, and in the southern counties of England farmers were encouraged to kill them as the 'regular limbs of Satan'. All over England they are known as the Devil's birds for their high-pitched calls, the sound of screaming souls departing to hell, but they send a wave of happiness and excitement through me because now I know summer is just around the corner.

Thursday 5 May, 2022 (low tide 1.12 m @ London Bridge, 11.17)
Central London – South Bank

A woman is trying to get her two children off the foreshore and into the Tate Modern art gallery beside the river at Bankside. 'Please, we'll miss our time slot,' she begs, 'we'll come back, I promise.' But her two little girls are having too much fun collecting old horse teeth and bones to worry about art. They dart across the foreshore, snatch up treasures, run away from waves and paddle about in the mud. They are filthy and happy. Eventually their mother gives in and agrees to a handful of morbid treasures. She cleans them up with a tissue from her handbag, and after a lot of complaining and promises, the girls reluctantly follow her up the steps.

Children have an innate curiosity that dulls with age. Everything is new and interesting when you're small, but all too soon other things take over and the simple things start to fade. We stop noticing. For some people this curiosity vanishes altogether, but for others it lies dormant and can be rekindled. It's fascinating to watch people search the foreshore for the first time. Some have a natural aptitude, while others become blinded and frustrated by effort, looking too hard and seeing nothing. Children, on the other hand, with their keen eyesight and colourful imaginations, can be remarkably good at finding. I search with my own children, on beaches and in fields, but I don't take them mudlarking. They think it's 'boring' and, selfishly, I'm quite pleased they don't want to join me in my sanctuary. It's my precious time alone, and anyway they don't give permits to anyone under twelve.

With the frustrated woman and her children gone, I turn my attention to an area that has been eroding nicely over the last year. I've found buttons and coins here and it has become one of my regular find spots. Today, it didn't disappoint. My first find is a small Charles I rose farthing, closely followed by a Charles II (r. 1660–85) jeton, a thin brass token that's tantalisingly half buried. On one side of it is a bewigged portrait of the king that gives away its age. This one has been made by a man called Cornelius Lauffer between 1660 and 1685 who, like most jeton makers at the time, worked in Nuremberg, Germany. The word jeton comes from the French *jeter*, to cast or throw. From as early as the twelfth century they were used for calculating at the Royal Exchequer, and by government officers and merchants for counting and accounts. When Arabic numerals (0, 1, 2, 3, 4, 5) replaced Roman numerals (I, II, III, IV, V) from around the fifteenth century, their traditional role became redundant and low-value jetons, like this one, began to be used as gambling chips in London's many gaming houses.

Just below the fruitful eroding patch, I find another counter in a thin line of floating debris and small bones.

118

It is a thin, flat bone disc, about the size of a 10p piece, and simply decorated with a line around the edge. I have found similar counters on other parts of the foreshore that have been identified by Stuart as Roman, but this one is different – thinner, plainer and a slightly different shape. Simple, handmade pieces like this are very hard to date and I suspect Stuart or Jo would probably slap a 'post-medieval' catch-all on it, but given its location, I think there is a good chance it could be sixteenth century. It may have been used in one of the taverns, brothels or gaming houses that made this stretch of the river so notorious.

I drop the game counter into my bag with the coins and go back to re-scan the patch I've already looked over. Sometimes a different angle, shadow or light can reveal finds that I miss on a first sweep, so I usually check over an area several times before moving on. It is just as well I do that today. Within yards of where I found the bone counter, I spy a small dull-grey circle barely breaking the surface of a puddle. 'Another cheap modern ring,' I think as I pick it up, but as I peer at it more closely, I see an interlocking pattern of circles, diamonds, lines and dots around the outside. It is thickly tarnished, cut right through on one side and twisted open, presumably to get it off. I adjust my glasses and angle it to the light. On the inside of the band is what I hoped to see – a letter 'A'. It is then I know it isn't modern.

As I turn the band slowly, more letters swim into view and my heart begins to pound. Reading them one by one I put together the word: 'ALWAYES'. It is a silver posey ring and from the style of the writing I guess it is probably about the same age as the first one I found about eleven years ago. My first posey ring is plain on the outside and has the rather sorrowful motto 'LIVE IN HOPE' engraved inside. I read everything I could find on posey rings back then, so I know that what I have in my hand was made and gifted as a token of love around the sixteenth century. The deep decoration cast around the outside may have been enamelled, probably in black or white, and it would have been a beautiful thing, until its owner took a pair of snippers and cut it off. But I think it is this damage that makes it so special.

The ring may have been cut off out of necessity, a swollen finger perhaps, but since it ended up in the river, I suspect there is more to this story than that. Sitting on my palm isn't just a damaged piece of sixteenth-century jewellery; it is a moment of grief, anger and regret, a simple symbolic act so profound that it has travelled through time without losing any of its potency.

Since it is made of silver and over 300 years old, it legally qualifies as Treasure, and I need to get it to Jo the FLO so that he can send it to the coroner within fourteen days. It will be offered to museums. If any are interested in acquiring it, there will be an official inquiry to declare it Treasure and the ring will be valued by an independent Treasure valuation committee. If none of the museums are interested, the Crown will disclaim its interest and it will be returned to me. Whatever happens, once I hand it over, I am not going to see it again for several years.

My luck continues as I wander along the waterline. I pick up a very worn eighteenth-century penny, made of tin to bolster the failing tin-mining industry, and a green-glazed ceramic bobble or 'knop' from a Tudor money box, which may or may not have been used to collect entrance fees in Bankside's playhouses and bull-baiting rings. In three hours, in an area less than a hundred feet square, I have reaped a bounty. Days like this are rare, but sometimes even areas that are well searched can throw up a trove... then on the next tide produce nothing at all.

Thursday 12 May 2022 (low tide 1.47 m @ London Bridge, 17.47)
Central London – North and South Bank

We moved out of London when the twins were three, and each year since, my birthday treat has been a night in a hotel close to the river, so that I can easily catch three tides in a row. I have two whole days with nothing else to do but doodle around by the Thames. Bliss.

I decide to start my first tide at Bankside and to swap over to the north bank halfway through. I buy a coffee and sit down to wait for the tide to fall. It is busy, so I start people-watching to kill time. I have a game I like to play in my head when I'm in crowds with nothing else to do. I've done it for years and it's entertained me on many delayed trains. I search the crowd for faces that don't fit, people whose features belong to another time. An older woman with a red face and squashed nose who would look far more at home under the cloth cap of a Georgian hawker, a gangly youth who should be hanging out in some dark Dickensian alleyway, a portly gentleman more suited to a doublet and hose and a little girl with the doll-like face of a World War II evacuee. Most people's faces will fit into another time, if you look at them carefully enough.

Fitting people who don't fit entertains me until the water is low enough for me to return to the patch where I found the posey ring the other day, and straight away find a lead token. Lead tokens have been produced for centuries and tend to echo the size of coins at the time they were made. Medieval tokens are usually quite small, and the larger size of this one suggests it is eighteenth century. It may have been issued by a local business to hand out in lieu of small change at a time when pennies and farthings were in short supply, or maybe as a tally token for counting cargo on and off barges and ships. The issuer's initials, IH or HI, are cast into one side. Was he Isaac or Henry, or perhaps Isabel or Harriet? Close to where I find it, I pick up a cufflink from around the same era, decorated with a dimple pattern a bit like those on a thimble, and a natural knot of wood that has been worn and sculpted by the river into an interesting shape. I carry on searching up to Gabriel's Wharf by the OXO Tower and do well. By the time I return to cross the Millennium Bridge, my finds bag is already full and heavy.

Halfway across the bridge I stop and look down. On both sides of the river, the foreshore is busy. On the south side it is mostly curious visitors wandering aimlessly and throwing stones into the water, but the north side is crowded with serious searchers. This part of the river has always been the busiest, especially on weekends and in good weather, but over the last two decades it has become steadily busier.

Bankside was little more than a building site until the late 1990s when the Globe Theatre opened, quickly followed by the Tate Modern and the Millennium Bridge. This brought crowds of people to what had been a fairly quiet, run-down and ignored part of London, and with them came concerns for the foreshore. But there were already concerns before the revival of Bankside.

In 1956, Ivor Noël Hume, who was introduced to mudlarking himself by a radio programme, wrote, 'Magazine articles, talks on the radio and television have all served to introduce the mudlark and his treasures to a wide public.' He goes on, 'Soon after the Second World War the river's trinket box was opened virtually for the first time, and everything that could readily be seen was taken by the many mudlarks who foraged *en masse*.' Geoff Egan, who pioneered liaison between archaeologists and mudlarks, followed this up in 1977 in an article for the *London Archaeologist*, where he wrote, 'This summer, the mudlarks were an all too obvious advertisement to curious visitors and resting workers of the prizes sought, and so the number of searchers was constantly increasing. They could also be encouraged by recent coverage of this activity in the media' – mudlarking had been filmed for the television show *Jim'll Fix It* in 1976.

By far the greatest reach and influence these days though has come from the ever-increasing number of people posting on social media and YouTube. Since I began posting on Facebook, anonymously at first, in 2012, mudlarking on social media has exploded and morphed into something very different. There are now hundreds of mudlarking pages and with some of them has come a kind of competitive mania for followers and 'likes' that I think is completely at odds with the peace and harmony of the river. I've even met people who say they aren't happy until they find something 'post-worthy', which I think is a shame.

Halfway down the river stairs under the bridge I see a wheelchair neatly folded up against the wall. Then I spot two familiar figures in the distance. They are sitting side by side on the shingle in the sunshine. The smaller of the two is a woman with dark bobbed hair and a blue jacket like mine, the other is a tall, willowy man with long, expressive hands.

Johnny Mudlark, aka Johnny the Bead, is back on his patch for the first time in years. He used to be a regular here, always searching the same small area, looking for beads and collecting anything else that caught his hawk-like eye. His last visit to the foreshore was in February 2020, after which lockdown and the progression of his functional neurological disorder (FND) made it difficult for him to return. He is sitting next to his old pal Liz, through whom I contact Johnny, who doesn't have a mobile phone or a computer.

The first time I met him, years ago, he showed me a tiny notebook in which he painted the objects he found, to size and in exquisite detail. They were stunning. Each one could have tumbled off the page and landed back in the mud where it came from. He generously let me use his paintings for the jacket of my first book and to illustrate my second book, and I'll always be indebted to him for that.

FND is a problem with the functioning of the nervous system, how the brain and body send and receive signals. 'I can still draw,' he tells me today, 'but having a shaky hand, limited concentration and double vision makes painting miniatures in my book an impossibility. I'm glad I painted them when I did, though.' I ask what he's doing instead. 'I make axe heads now, in my kitchen,' he says, 'from various found rocks; some of them are from the foreshore. And I still have my mudlarking collection. I made two shelving units for them out of an old dumped bed. I filled the shelves with various-sized labelled jars of mudlarking joy. Jars of London Preserve in a museum larder!'

We sit and chat and reminisce until the tide comes in, then we walk slowly back to the river stairs. I help them get the wheelchair to the top of the steps, where I say goodbye. I've missed Johnny.

Friday 13 May (am) 2022 (low tide 0.83 m @ London Bridge, 06.19)
Wapping

I am out of the hotel by 4.15 a.m. and on my way to Wapping. The hour before dawn is a special time in the city. You can walk down the middle of quieter roads and on the main streets the only traffic is the odd delivery van or lorry. I pass four men looking a little worse for wear. I guess they have been on a bender. An early jogger thuds past a man with a suitcase who looks lost, and there's a sad-looking couple walking hand in hand. Shift workers and security guards are on their way home and cleaners are heading to offices in the City that never switch off their lights.

I cut down a path past the Tower of London where the sweet, bubbling song of an early wren bursts out of the bushes and fills the silence. Little Jenny Wren, who despite his size outwitted the eagle by hitching a ride on his back and springing off to fly higher and claim the title 'King of Birds'. The mellow flutey notes of a blackbird join in and echo off the ancient stone walls. Blackbirds sometimes sing all night in the city where the lights from offices and street lights confuse them.

Dawn, daybreak, sunrise, cockcrow, sun-up, dayspring – on good days this is my favourite time. The day is mine and it is filled with possibility. On other days, the hour before dawn can feel like an eternity, the loneliest hour of the day. There is a big difference between being lonely and being alone and it is entirely possibly to be both alone and lonely in a room filled with people. This morning I am truly alone, there is no one around and it feels wonderful.

My first glimpse of the river is across a small park. It is already low, calm and flat, a huge blindworm inching slowly east. In the distance, behind the black masts of the antique sailing-barges-turned-floating-homes at Hermitage Wharf, the sun is just below the horizon and the sky is pink. I usually escape from the traffic and cut off the road around here, onto the river path, but it is so quiet I carry on. Without the sounds of cars, I can hear pigeons, starlings, crows and even the wind in the trees. I stop for a moment and close my eyes. I could be standing on a country lane. River sounds trickle down the little alleyways that lead to river stairs and the only car, an electric taxi, catches me by surprise as it whirrs past.

The cobbled streets are wet and shine like polished leather. Even at this time in the morning they would once have been loud and busy, filled with dockworkers, warehousemen and sailors, but this morning they are empty. Just a squirrel dashes over the road and scrambles up the wide trunk of an old lime tree. I look up and see bedroom lights in houses and apartments flicking on as the street lights turn off. The sudden darkness stirs a fox, who appears from behind some bins and slinks across my path, casting me a sideways look before vanishing through some black iron railings and into a playground. It leaves behind the distinct musky smell of dog fox and a torn bag of rubbish.

I pass the Town of Ramsgate pub, where there is a soft light behind the bar and towels neatly draped over the beer-pump handles, and turn just before Wapping police station into a small park. The park is unlit and it makes me nervous, so I hurry through it to a short flight of concrete stairs up to an iron gate that leads to a ladder onto the foreshore, where I am safe. The approaching sunrise is turning the sky violet and orange, and the river is a dark mirror beneath it. The Clippers haven't started yet, which means they haven't washed away the natural layer of thin silt, as soft and grey as mouse's fur, that's usually left behind by an undisturbed tide. I need the first Clipper to send its wash across the foreshore before I can start to search properly, so I crunch and slide over the muddy shingle to a good spot to watch the sunrise. I find a slab of muddy concrete large enough to make a comfortable seat, get a plastic bag out of my rucksack to sit on, pull my jacket up around my ears and wait, and think.

Three ducks float motionless on the water between me and the sunrise and the foreshore is silent. London is holding its breath for the new day, and then it comes. The sun appears quite suddenly over a low line of buildings far away in the east and rises surprisingly quickly until it is a poker-red eye in the violet and orange sky. The soft orange light of the sun picks out the wooden posts and lumps of concrete on the foreshore and reflects off the wet mud, making it greyer and bleaker. It is a strange apocalyptic scene in stark relief against the beautiful sky.

As the sun pulls itself up in the east, I hear a deep rumble in the earth beneath me. The East London Line is running, and, in the distance, the muffled sound of traffic tells me the city is waking up. The first Clipper washes past and suddenly there is pandemonium. Two drakes start to fight over a duck; a murder of crows stretches up, leans back and shouts at the sky; and an Egyptian goose finds its way to the top of a mooring pole, where it honks loudly at everything and everyone. In the shallows a coot flaps angrily out of the water at a crow, and screaming gulls wheel in the brightening sky over the whole chaotic theatre.

I leave my thoughts on the slab of concrete and walk towards the kiln dumps. I like to keep an eye on how fast they are eroding and if anything new has appeared. I've found four of them so far at Wapping – salt-glazed stoneware, delft, transferware and the remains of a pipe kiln. I think they all date to around the mid-eighteenth to early nineteenth century and may represent a call-out to local industries for hardcore and waste to stabilise the mud and fill in holes on the foreshore. The stoneware waste consists of broken pots, vase-like saggars that supported the pots in the kiln and allowed heat to circulate around them, and random squidges of clay that were used as separators and supports. I collect these hastily made sausages, blobs and columns that were indirectly glazed when salt was thrown into the hot kiln to glaze the pots. They often have finger, thumb and palm prints baked into them that I can fit my own hands into. It is an eerie feeling, like shaking hands with the past.

The delft kiln was cleared out at biscuit firing stage, before the vivid blue and white colours were added. The shards are porous and plain yellow, and the nose-like kiln stilts that would have supported the bowls and plates as they were being fired are mixed up among them. The transferware waste is burnt and bubbled, and sometimes fused together. Whatever happened in that kiln was a disaster, and the ruined pots would have been known as 'pitchers', since they had to be pitched out and dumped.

I had previously thought all that was left of the pipe kiln were little sections of broken white-clay pipe stems and the occasional complete pipe bowl, but I've been told there are also pieces of the pipe kiln itself, and that is what I want to look for today.

I begin to search where the Clippers have washed away the silt. I find an early twentieth-century glass sauce-bottle stopper, a couple of plain nineteenth-century trouser buttons and a tiny green seed bead, which is impossible to date. Unless you can find a dateable comparison or they are found in context with other dateable objects, most beads are almost impossible to date accurately. Then I pick up something that I know is infinitely older, a black heart-shaped flint. It is a micraster, a type of sea-urchin fossil, that will increase my collection of foreshore-found flint sea urchins to sixteen.

At the site of the pipe kiln, I collect three complete pipes and a broken decorated pipe. Then, at the water's edge, I see what looks like a lump of concrete with pipe stems embedded in it. It's what I've been looking for: a muffle fragment from a pipe kiln, made from raw pipe clay moulded around previously fired but broken clay pipe stems that were used as reinforcing bars, like wattle and daub. The muffle was a chamber inside the kiln that resembled a giant pot, within which the stack of pipes were placed to keep them separated from the kiln gases so that they would stay white during the firing. The muffle chamber remained in the kiln and was reused again and again, so broken pieces of it are quite rare. It isn't a thing of beauty, but it is worth adding to my collection, so I bag it and carry on.

By the time the tide ushers me off the foreshore, the sun is high, and the sky is blue. I retrace my steps back through the park and past the playground, where somewhere a large dog fox is dozing in the sunshine, along roads that are now busy and noisy, and back to my hotel room, where I shower and pack.

Friday 13 May (pm) 2022 (low tide 1.11 m @ London Bridge, 18.38)
Custom House

I meet a friend for lunch and return to the foreshore in the afternoon for one more quick look before I go home. I decide on Custom House, which I haven't been to for a while. It is also conveniently close to the hotel for picking up my bags afterwards.

Custom House is a fairly short stretch of foreshore with a long and complex past. It is on the north bank, just west of the Tower of London, and close to where the first permanent bridge over the Thames is thought to have been built by the Romans. It is also on the original Pool of London, a naturally deep part of the river that has accommodated seagoing vessels since at least medieval times. The name Custom House comes from the custom houses that have been here since the fourteenth century, where duties on imported and exported goods were paid. It is a powerful and ancient part of the river and I wonder how much it has hoarded away in its depths.

At the westerly end of the Custom House foreshore is old Billingsgate Market. There is thought to have been a market here since Roman times, but the first written record of Billingsgate dates to the toll charges of 1016, which charged a halfpenny for a small ship and a penny for a larger one with sails. At the time, the city's main fish market was near Queenhithe Dock, but as the boats grew larger, they became unable to sail under Old London Bridge and took to unloading at Billingsgate Dock instead. By the sixteenth century it had become London's main fish market. The fish was unloaded from boats that tied up along the quayside and sold at wooden sheds and makeshift stalls around the dock. The waste that built up around the market stalls was periodically shovelled into the river, where it was sucked down into the mud and preserved.

Shouting, swearing, bartering and the stench of rotting fish guts would also have tumbled down onto the foreshore from the market, where pickpockets slithered through the crowd like eels and drunks lurched from tavern to market and back again. The traders were rough and hardworking, and the women who gutted the fish, the Billingsgate fishwives, were infamous for their willingness to fight. They worked long, filthy hours in atrocious conditions with hands that cracked and split in the wet and cold, and cuts and chilblains that turned to weeping sores that never healed. Their dresses and aprons were stiff and reeking from fish blood and guts and their language was so crude that the word 'Billingsgate' became a byword for swearing.

I wonder how many of the clay pipes I've found here over the years fell from between the rotten teeth of a Billingsgate fishwife.

Eventually a decision was made to clean up and develop the market site, and in 1876 the market building that still fronts the river was opened. It was a cathedral to fish, with a huge ice room in the basement and a large central arcade where the fish was sold. Ships and fishing boats jostled in the water in front of it waiting to unload, merchants haggled and porters, wearing special tarred leather hats with brims to stop the fish juices from dripping down onto them, rushed to and fro with teetering boxes balanced on their heads. They kept a tally of their loads with specially made tokens that sometimes ended up in the river.

With all this activity you'd expect there to be a lot to find at Custom House, and there has been. The foreshore was well dug over by mudlarks in the 1980s, when the old Billingsgate Market car park was demolished and a new walkway constructed. It produced swords, medieval pilgrim badges, seventeenth-century pewter toys, coins and a lot more, but I've never been that lucky here. Still, it's usually a quiet patch and the perfect place to while away a few hours before my train.

I begin under Sugar Quay Jetty, where there's a tangle of modern iron and junk in the mud and sometimes something more interesting in among it. I find a sixpence dated 1947, the year the Cold War began, and work my way west, searching behind the revetments closer to the water, where I reason things must have washed up. Sure enough, there is a line of pins, nails, blobs of lead and other small metal pieces where the shingle meets the sand, and in it I find a large handmade copper-alloy pin with a satisfyingly large head. I also find a late nineteenth-century trouser button with the name and address of the tailor, 'M Isaacs, 108 Hoxton St.', around the edge of it.

Tailor buttons are fun and fairly easy to research through business directories and censuses. I find Moss Isaacs listed as a waistcoat maker in the 1861 census and a 'clothier' in Kelly's Directory of 1891. Moss and his father Elias were both born in London, but his grandfather Barnet was from Poland and, like many, is likely to have fled antisemitism for London's East End. I find Moss's workshop on Google Maps: 108 Hoxton St is now a posh falafel shop next to a French-style boulangerie. I wonder what he would have made of that.

By the river wall, the sand is a coarse grit made from shells that were dumped into the river from the market. There are more oyster shells here than anywhere else on the foreshore, along with crab claws and the shells of scallops, mussels, cockles and thousands of winkles. The winkle shells have survived better than the others, probably because they are thicker and more robust.

I cross the skeletal wooden frame of a gridiron, where ships once rested at low tide, and plunge into the gloom beneath the walkway in front of Old Billingsgate. It is dark and muddy under the concrete path, water drips and dribbles from above, and it is slow going. I climb over and between metal struts and concrete pillars, clamber and toil over piles of rubble and splatter through deep puddles of mud, but I really need a torch to see anything properly and I don't have one with me. The only thing

I find is a fragment of nineteenth-century clay pipe bowl with the name of the maker, 'Benson' (George Benson of Pentonville 1795–1825) stamped into it. I could walk much further into the slimy tunnel, but I am tired, and I decide enough is enough. I make my way slowly back to the stone steps near Sugar Quay, muddy, happy and one year older.*

Tuesday 17 May 2022
The Bellarmine Museum

Alex is something of a legend in the mudlarking community and a visit to his museum in Norfolk is a pilgrimage many have made. I've been trying to make time to visit for years, and today I finally make it. Alex is the fount of all knowledge when it comes to post-medieval German stoneware ceramics with wonderful Tolkienesque names like Westerwald, Siegburg, Frechen, Bouffioulx and Raeren, but his speciality are Bellarmines. These are pot-bellied flagons with bearded faces on their necks, and medallions – usually the coats of arms of the towns where they were made or the merchants who commissioned them – stylised flowers and merchant's marks on their bellies. It is these I really want to see.

In Germany, where most of them were made, they were called Bartmann (bearded man). They were also known as baardman, bartman, bartmankrug and baardmannzeug, but in England they became known as Bellarmines to poke fun at the Catholic Cardinal Bellarmine, who was staunchly anti-alcohol and on the wrong side of the Protestant/

---

* Please note, mudlarking is no longer permitted on the Custom House foreshore

Catholic divide in the sixteenth and early seventeenth centuries. It is said people smashed the bottles to see the cardinal's face in pieces, but that's probably just urban myth. The real reason so many of these mottled shards show up in fields, gardens and rivers is because they were so common. I have a drawer of pieces, mostly just an eye, a grimace, a piece of tangled beard or a medallion, but it is the complete faces, of which I have quite a few, that are the holy grail for mudlarks.

My friend Emma, who lives in Norfolk, is meeting me at the museum and pulls into Alex's drive soon after I arrive. She is a 'sometimes mudlark' and the last time we went to the foreshore together we found a beautiful Bellarmine face, so she really had to come. Alex is waiting for us at the back door in a purple tie-dye T-shirt with Alice in Wonderland's Cheshire Cat grinning out over his belly. 'Do you like my T-shirt?' he beams, and shakes our hands. He has long white hair and a beard, albeit short and stubbly, and it wouldn't be unkind to suggest he looks similar to the bottles he collects. 'Come in and see what I picked up yesterday,' he says excitedly, and ushers us straight through the kitchen and into the house.

It is an Aladdin's cave of curiosities: fossils, bayonets, cannonballs and swords, delft plates and tiles, cabinets filled with all manner of miscellanea, collections of clocks and ginger pots. I notice an old oak coffer similar to mine in each room (his partner Deanna wouldn't let him have any more), and shelves and mantelpieces, including special high shelves close to the ceiling, are filled with pots, flagons, tankards and jugs. Most of them are shades of salt-glazed speckled brown and grey, but in among them are busy bright cobalt blue patterns and splashes of purple manganese. 'I collect everything really, but mainly stoneware these days, I outgrew my last house, so we had to move for more space, but I've outgrown this one too.' I can see that he has, and it makes me feel a lot better about my own collecting.

The new acquisitions he wants to show us are in the sitting room. Nestled on the sofa, spilling out of plastic crates and arranged in groups on the floor is a mob of Bellarmines, if that's what the collective noun would be. 'It's a private collection,' he says rubbing his hands together in glee. 'I've owned some of them in the past, so they've come back to me, but there are bottles in here I've never seen before.' He picks one up and gazes at it lovingly. 'I couldn't sleep last night for thinking about them, but Deanna isn't too happy about me buying more.'

A hundred tiny stoneware eyes stare at us, and one of them winks at me. It is a much smaller bottle, only about 5.5 in high, with a very gentle and detailed face. I pick it up and run my finger over the touch point on its belly, a 500-year-old scar where it was stacked against another bottle in the kiln. It had never occurred to me they came this small. 'Oh yes,' says Alex. 'That one was probably for mercury, it's from about 1580.' Mercury can smash a glass bottle if it's shaken, so a nice thick stoneware pot like this would have been perfect. 'I'm letting it go if you want to buy it,' says Alex. 'I'm sorting out the ones I think I can sell, and the rest will go in the collection. Do you want to see it now?'

I reluctantly put the little pot down and we walk slowly back through the house and outside to a converted garage.

'You wouldn't believe the security I had to install to get them insured,' Alex says over his shoulder as we make our way up the path. He unlocks it, flicks on a switch and my paltry collection of thirty-three complete foreshore-found Bellarmine faces pale into insignificance in an instant.

The room is lined with Bellarmines of every size and style, dating from the sixteenth century to modern copies. In the centre of the room is a low, solid-mahogany museum cabinet that fills the length of the museum. 'That came from the Victoria and Albert Museum when they were having a clear-out,' he says. It is filled with shards of faces and cartouches, and even the specimen drawers underneath are stuffed with broken bits.

We start at the edge of the room and work our way slowly around. Alex explains where each one came from and how he tracks them down: 'I've got about 400 stoneware Bellarmines at the moment, but of the thousands I've seen over the years, I've only ever found two identical ones,' he says. And I can believe it, they are as individual as we are.

Some of the seventeenth-century faces are more crude and ugly but the earlier Bellarmines, dating from the sixteenth century, are finely crafted, with beautiful flowing beards. Some of the beards are arranged in patterns, some are tied up with string or ribbons and others are so stylised they barely look like beards at all. I look carefully at each face and see ears of corn in their eyebrows, berries and flowers in their mouths, leaves hidden in their beards, and noses that look like phalluses. It all hints at the bartmann's origins as the wild man of European folklore and his connection with fertility and the land.

My favourite group is the wonky ones. They had collapsed or sagged under the weight of a stack of pots in the kiln before they were fired, and were probably sold locally as seconds. They may not have been perfect, but effort had gone into making them and they still did the job they were made to do. 'Earlier ones have wider mouths – they doubled as tankards for people to drink from, but the later ones were for storing and pouring,' Alex explains. He's found tar in some of the larger bottles, and they were probably also used for storing acid, oils and vinegar.

Bellarmines held protective spells too, and have been found containing urine, nail clippings, iron nails, pins and human hair. They were buried under doorsteps and near fireplaces to prevent malevolent spirits from entering the house. They are rare, but Alex has six. 'One was in an antiques shop near here for twenty years before the owner agreed to sell it to me; another was on eBay, cheap because the top was broken.' He found nails inside that one and emailed the lady who sold it to him to ask what she knew about it. She explained that she'd found it under her granny's doorstep and had broken the top trying to open it. When she did, she said it smelled of stale urine.

Hours pass in the garage, and we are ready for tea by the time Alex announces, 'Cakes!' 'I'm not supposed to have cake, I'm diabetic, but it's OK if I have guests,' he says with a guilty look. We tuck into date slices and strawberry tarts piled high with cream as he tells us more about his collection and how he got started. 'I was on my way to work, and I broke down. This chap stopped to give me a lift and he had a metal detector in the back of the car. He showed me a coin he had found and offered to take me to a field where there was Roman pottery everywhere. That's where I got the bug for field walking, then I started digging and searching building sites.'

Alex found his first Bellarmine in 1976. It was embedded in the side of a large pit in what had been the backyard of an old pub, and it captured his imagination. 'I imagined the chaotic scene in the tavern where you could get drunk for a penny and dead drunk for tuppence, and carried it home wondering how it had ended its useful life. Now I buy them from auctions, dealers and other collectors. eBay's no good any more and car-boot sales aren't what they used to be,' he says. One of his best finds was from the good old days of car-boot sales though: two Bellarmines, painted gold and turned into lamp stands. He paid £4 for the two and dropped them into a tank of caustic soda to get the paint off.

'Are you a collector or a hoarder, Alex?' I ask.

'I'm a hoarder – all collectors end up hoarders, there's no difference. Luckily, I've got Deanna. She calms me down, she knows what's in the house and if I try to bring in another one, she'll put her foot down. Even if I try to hide them. Anyway, it's all for sale,' he says laughing. 'We're only looking after them for the next generation. I've recorded everything, it's all been published, so I'm happy to let them go.'

When I had arranged to visit Alex, he had warned me to make time for it: 'Most people stay between two and six hours.' I'd laughed, expecting to stay two hours tops, but five hours has flown by, and it is time to go. As I drive away, I make a decision. I don't think I'll ever find a complete Bellarmine on the foreshore, so I decide to email Alex when

I get home to tell him I'll buy the little mercury Bellarmine that winked at me from the sofa.

Tuesday 24 May 2022 (low tide 1.59 m @ London Bridge, 14.19)
Central London – North and South Banks

Just as I pick up the rusty head of a small hammer, a strange yellow light creeps over the foreshore, illuminating every grain of sand and pebble, and turning the river bright green. It is a day of curious colours. Low, grey, thunderous clouds roll in from the west and a wall of rain sweeps downriver. I feel the pressure lift as it draws closer, a breeze blows up from the river, then the first fat drops fall.

The torrential rain turns to sheets of hail that bounce off my jacket and sting my nose and cheeks. I run for the river wall, my only cover, and press myself as close to it as I can. Through the noise of the thunder and hail I hear people shouting on the path above me and footsteps running for shelter. I look across the river. The opposite bank is barely visible, and the water in between is crazed and shattered by the icy bullets. It builds up on the foreshore in clean, white frosty piles and I think of the twins last summer, laughing as they danced in a long-awaited rainstorm. I don't know why I didn't join them, but afterwards I wished I had.

The storm sweeps through quickly and ends as abruptly as it had begun. It takes the curious light with it and leaves behind a strange post-storm silence that for a moment is filled with the mellow song of a blackbird.

Thursday 26 May 2022 (low tide 1.17 m @ London Bridge, 16.39)
Central London – South Bank

A man at the top of the river stairs is watching a seagull pull the well-pecked corpse of a pigeon out of the shallows. A crow is standing by, patiently waiting for his share. As soon as the gull has done the hard work, the crow shoulders it out of the way with an angry squawk and begins to peck at the remains hungrily. There is always at least one crow on the foreshore in central London and I have watched their numbers rise steadily over the years. Some of them are quite tame – I think other mudlarks must feed them – and they follow close to my boots as I crawl over the rubble and mud. They peck at the freshwater shrimp I uncover and mudlark for themselves, turning over stones and roof tiles to get to the creatures underneath.

The crow regards me with a bead-like eye, a lump of pink flesh stuck to its shiny black beak. It is both repulsive and compelling and, like the man at the top of the stairs, I find it hard to look away. They are friendly yet sinister birds. Dark, stalking figures, descendants of the crows that picked flesh from gibbeted bodies that were strung up beside the river further east and the heads that were stuck on long poles at the gateway to the southern end of Old London Bridge. If I had been here in 1305 I might have just been able to see the head of William Wallace (Scots rebel or freedom fighter, depending which side of the border you were from) in the distance, his mouth pulled open by the pole that was rammed through it and his long hair waving in the wind. The Keeper of the Heads was in charge of putting up new heads and of taking down the old ones. I wonder how many he threw in the river?

The sun has brought people out today and the river is busy. A man is sitting in the middle of the foreshore on a dirty white-plastic garden chair that was left behind by the last tide. He is staring out sadly across the river, an empty litre bottle of cider next to him and the next one already half finished. People come to the river for all reasons and commune in their own way.

I try not to disturb him, but it's hard not to clatter on the flint cobbles as I tiptoe past. He doesn't seem to notice, though, and soon I'm on softer sand, heading west. I pause to watch a honeybee supping from a river puddle and I'm wondering if it has come down from the hives on top of the Tate Modern, when I spot a Charles I silver penny stuck to the sand. The mint mark tells me it was struck at the Tower of London and it reminds me of something I had been reading earlier in the week about conditions at the Mint. The dark, hot workshops where the coins were made were filled with toxic fumes. In 1560 several German workers fell seriously ill, probably from breathing in the noxious gas. They were advised by other workers at the Mint to drink milk from a human skull, but despite this, several of them died.

A little further on I find a less deadly coin, an American quarter dated 1996 and, in a pile of old nails, an iron-punch, possibly for leather working, and the flattened bowl of an eighteenth-century rat-tail pewter spoon. I also find another jeton, earlier than the one I found a few weeks ago. This one was struck in Nuremburg sometime between 1570 and 1613 and, like many jetons of the time, it has a moral warning written in High German around the outside. It reads 'GOTES GABEN SOL MAN LOB', 'One should praise God's gifts', and while I'm not religious, I take it as a gentle reminder from the river to be thankful for what I have.

SUMMER

CODE: 216.22.GW02

OBJECT:	18th century apothecary bottle
MATERIAL:	Glass
DATE FOUND:	21/06/2022
LOCATION:	Central London – south bank
NOTES:	A small (5.5 cm) clear glass, free blown, cylindrical bottle with a short neck, a flared lip and kick-up base. The inside is stained dark brown and smells of tar. Small bottles like this were used to store, dispense and transport oils, powders and chemicals. Partially buried.

JUNE

Central London – South Bank

I walk down Cardinal Cap Alley for the first time this afternoon and through a tiny wooden door into the garden of No. 49. The alley is the last remaining Tudor street on Bankside and No. 49 is the house by the river, next to the Globe Theatre, that everyone walks past and says or thinks, 'Imagine living there.' I look up at it and think, 'Imagine the tides it's seen.'

Built around the year St Paul's Cathedral was completed, for 300 years it has watched the river flow, seen wondrous water pageants pass by and witnessed frost fairs that stilled the water and brought the city down to play on its ice. It has seen bridges built and demolished, watched the City of London across the water cower under the Blitz and then rise again as a forest of glass and steel. Wherrymen have come and gone, ships have turned from oars and sail to engines, and the river has been poisoned with pollution and reborn. The only constant and predictable thing this house has seen in its time is the tides that will continue to rise and fall well after No. 49 is rubble and its dust has blown into the river.

The garden is a green oasis on a hot day. As wide as the house, it is cool and shady and smells of jasmine. Stepping back out into the frenetic crowds on Bankside is a sudden jolt back into reality. I cross the line in the cobbles where the riverside used to be before it was pushed north. The Ordnance Survey map of 1869–80 shows a 'High Water Mark of Ordinary Tides' that came much closer to the house when it was surrounded by wharfs and yards and a collection of other tightly packed dwellings.

Bankside is busy and the foreshore is hot after the calm and cool of the garden. At the top of the foreshore, in the shade by the river wall, a man has set up turntables and speakers and is DJing to no one in particular. Maybe he's playing music for the river. The tide has already turned, and the mud has been well picked over by other mudlarks. The river has reorganised itself since my last visit and thrown a blanket of shingle over my special spot where I'd found my posey ring in May. The Thames is an unpredictable creature that can leave the most delicate objects alone for months and then suddenly move a giant timber or large stone on one tide. Even if I revealed the exact location of a good spot, it is likely to have changed by the time this book is published.

I head west to the eroding bargebeds near Gabriel's Wharf. There's hardly anything left of them now, a combination of boat wash, natural river action and people scraping and digging into them has worn them away. Bargebeds, or 'hards', are one reason mudlarks find so much. In its natural state the river is a gentle 'V' shape, but in the eighteenth and nineteenth centuries bargebeds were constructed to provide an artificial surface for flat-bottomed barges to rest on at low tide. A strong wooden plank-and-pile wall was constructed several yards from the river wall, into which was poured the city's rubbish, industrial waste, building spoil and anything else that was to hand, then it was rammed down and capped with a hard surface, often chalk.

Constant use meant that the bargebeds had to be regularly maintained. Through the 1920s and into the 1940s, two men, George Williams and Albert Hodgeson, were employed as 'foreshoremen' to push the sludge off the bargebeds at low tide to prevent the barges from sliding back into the river. Their job was so important that they worked by the light of the moon through the war years, so they didn't attract the attention of enemy bombers. But in the 1960s and 1970s, when the warehouses closed and ships and barges stopped coming into central London, the bargebeds were abandoned, and the river continued its work unhindered. It broke through the revetments, drilled down through chalk caps and scooped out the contents in its quest to return the riverbed to a natural V shape.

The early 1970s heralded another change. Soon after the barges and ships left, treasure hunters arrived en masse with shovels, pickaxes and metal detectors. For the first time mudlarks weren't just collecting what they found on the surface, they were also tearing up the foreshore in search of its hidden treasures. As early as 1956, Ivor Noël Hume had cautioned against this, saying, 'The Port of London Authority and wharfingers take great care to ensure the surface of the foreshore is not disturbed, for if the skin is damaged the less stable mud below can be quickly eroded.' But by 1977 Geoff Egan was comparing the scenes of destruction he saw to a war zone. 'In walking along the foreshore in the city this summer, one could come across areas of intense activity, which, when abandoned by the diggers, resembled the conditions of the Somme.'

Most active mudlarks are too young to remember those days, but they have been described to me by someone who was there as a 'Wild West and a free-for-all', with fights over territory and dealers lurking around the holes to buy objects fresh from the mud. Some important pieces were sold to the Museum of London, but it's impossible to know how much was sold privately or smashed to bits in the frantic hunt.

Ralph Merrifield, senior keeper at the Museum of London, wrote about his concerns in a letter to *The Times* in 1976:

There is a fundamental difference between harmless surface mudlarking and the digging of deep holes in the Thames foreshore by treasure hunters ... we now know archaeologically significant layers extend into the present river-bed, so that interference with them by unscientific digging is destruction of valuable evidence – as much to be condemned on the foreshore as on dry land.

In 1977, Geoff Egan also called for digging to stop, saying it would 'discourage those who only make financial profit, since it is medieval metal above all that seems to spell money to these: with the lower, earlier levels unavailable, few would consider the later and more scattered surface finds worth the bother of hours of searching'.

Eventually, the PLA and other agencies threatened to ban mudlarking altogether to protect the foreshore and its antiquities, but after negotiations, it was decided in 1979 to issue permits to those who agreed to abide by a set of rules. Under the first permit terms, 'Every object found by the holder, its exact location and details of the circumstances of its finding, must be reported as quickly as possible to the

Museum [of London].' Mudlarks were tacitly allowed to keep their finds, but the PLA retained legal ownership and to this day it is still illegal for mudlarks to sell what they find on the foreshore because they simply don't own it. They no longer receive a 'finder's reward' (half the object's value) either, which usually goes to the finders of objects found elsewhere. With digging officially sanctioned throughout the 1980s and into the 1990s, the foreshore in central London, especially on the north bank, was dug over like a potato patch.

In 1980, those with the new permit began a club called the Society of Thames Mudlarks, which is still in existence today. There are around fifty members, with a long waiting list and a one in, one out policy. Anyone with a 'standard' permit can apply to join as long as they have been reporting their finds to the Museum of London for at least two years and are willing to wait long enough to get in. The society meets monthly and has traditionally been quite secretive, but more recently they have been told by those in charge of the permit system that they need to be more transparent about their membership and more open about their finds.

As far as I know, the Society of Thames Mudlarks still has a predominantly male membership and many of those who are active on the foreshore are metal detectorists. Members have described their club as 'exclusive', which I suppose it is because it has limited membership, but I don't agree with those who describe themselves as the only 'real mudlarks', since historically most real mudlarks were women and children. I'm not a member of the Society of Mudlarks and I'm not on their waiting list.

With membership of the Society of Thames Mudlarks comes privileges: a 'mudlarking permit' that currently entitles members to dig to a depth of 4 ft in certain locations and to scrape and detect on the foreshore in areas that standard permit holders are not allowed to disturb. Anyone can apply for a standard permit, which allows metal detecting and scraping in permitted areas to a depth of 3 in, but although it is better protected than most on-shore land, the Thames foreshore is still being irreparably damaged by some searchers.

There are only a couple of society members who still dig deep holes today, but I've seen them going deeper than they should and digging where they shouldn't, and the damage they cause to this fragile environment can be significant. Even if they backfill their holes it doesn't stop the mud from eroding faster. Once it is destabilised, the river and wash from passing boats makes short work of softened mud. But of equal importance is the increasing number of mudlarks scraping with trowels and handheld rakes. Apart from the environmental impact – removing the top few important inches where invertebrates live and fish spawn – it is gradually peeling away layer upon layer of foreshore and it is one of Stuart's main concerns: 'People scrape much deeper than they should, then you get several people repeatedly scraping the same spot and there's a lot being lost.' He also blames the rise and abundance of cheap metal detectors: 'Detectorists dig lots of small holes, they rarely fill them in, and this is breaking up the foreshore. It's a long-term problem and it's not sustainable.'

Those who dig and scrape today vehemently and sometimes quite aggressively deny it does any damage, but over twenty years I have witnessed the long-term effects with my own eyes. I'd like to think we've moved beyond the kind of Victorian smash and grab that emptied Egyptian tombs and levelled prehistoric barrows, and it would be nice if, sooner rather than later, the importance of the foreshore and its contents was properly recognised and protected from indiscriminate spades, forks, trowels and metal detectors.

I kneel down to get a closer look at the crumbling edge of the bargebed. Someone has been scraping at the bottom and dug away quite a hole, but they haven't looked further up, where the thin edge of a grey semicircle is sticking out of the dark gritty mud. I pinch it between my thumb and forefinger and gently pull it out. It is another lead token. I've found a lot of eighteenth-century lead tokens over the years. Most of them have random patterns, cross-hatching and cross-and-pellet designs, but this is the first I've found with a date. It is written in the sloping style of the day with an elongated upper stroke to the last digit that could have made it an '8', but I decide is probably a '6'. When I get home I compare it to a farthing from around the same date and it is almost exactly the same size, proving the theory that lead tokens probably can be dated by their size.

Before I stand, I pick up some of the clods that have fallen out of the side of the bargebed and push them into the hole at the bottom. I know I am fighting a losing battle, but it is not one I am willing to give up on easily.

Thursday 16 June 2022 (low tide 0.58 m @ London Bridge, 09.06)
Central London – North and South Bank

The weather forecast has predicted a hot day and I'm pleased to make an early start. The air is cool as I walk from the station to the river, which is as calm as a lake with a whisper of mist hanging over it. The river path is quiet and in front of the Anchor it smells of the beer spilled by last night's outside drinkers. I keep walking to the Globe Theatre and at around 6.30 a.m. I jump off the end of the concrete stairs in front of it, and onto the foreshore with a glassy crunch.

The stairs used to be level with the mud, but each year the drop gets a little longer and it now hangs almost three feet above it. My friend Fiona, an archaeologist with thirty years of experience on the foreshore, remembers when the mud gave way to the chalk cap of an old bargebed. When that washed away, sacks of Victorian bottles appeared that must have been used as part of the bargebed's base. According to Fiona there was almost every kind of bottle you could imagine: tiny inks, heavy-bottomed beers and delicate perfume bottles. The sacks have mostly vanished now – just occasionally a strip of hessian will appear from the mud to flap and wave in the tide – and the bottles are all smashed. Now there is a layer of broken glass I wouldn't walk over in anything but a stout-soled pair of boots.

A while back I wanted to know where the bottles might have come from, so I did some research. The Anchor Brewery was founded nearby in 1616, close to the river for a plentiful supply of water. Its tap room was the Anchor pub, where Pepys watched the Great Fire engulf the City of London opposite in 1666. It has been expanded and partly rebuilt over the years, but there is enough of the original tavern left for it to claim to be the sole survivor of the many taverns and inns that lined the riverside at Bankside in the seventeenth century.

The brewery at Southwark was acquired by Barclay Perkins & Co. in 1781, and by the early nineteenth century it had become the largest brewery in the world, covering around fourteen acres of Southwark and attracting visitors from far and wide. It was said it used so much water that when the pump of the well at Barclay's was in use, the water level in the well of the City of London Brewery Company on the north side of the river fell.

It's possible the bottles came from the giant brewery, but there was also a large glass-making industry on Bankside that may have contributed. Noxious industries, such as tanning, dyeing and slaughterhouses, and those with a high fire risk, were banned from the City of London, so Southwark became a centre for glass-making. Glasshouses are recorded on Bankside from at least the seventeenth century until the end of the nineteenth century.

If the glassworks didn't contribute to the sacks of bottles under the bargebeds, they were certainly responsible for the smooth iridescent blobs and trickles of dark, cooled, molten glass that I often find at Gabriel's Wharf. They look like soap bubbles caught in the mud, reflecting a rainbow of colours in their dying surfaces because, like us, glass gets sick with age.

'Sick glass' is a term used for the chemical decomposition of glass over time. While ceramics remain largely unaltered when they are buried in wet conditions, glass begins to decay, resulting in discolouration, clouding, cracking and, eventually, flaking. One beautiful side-effect of all this is the iridescent hue it takes on as it starts to suffer. My collection of drips and globs is slowly dying and will one day flake away completely.

I've got myself into a rhythm of starting on the south shore at Bankside and heading north to see the end of the tide at Trig Lane. It is partly to avoid the crowds on the north side, but now that summer is here, the south is also shadier in hot weather and today it is hot. By 8 a.m. the sun is beating down on my back and my shirt under my rucksack is soaked with sweat. I have a battered straw hat for this kind of weather that's supposed to roll up, but it isn't faring well in my rucksack. It's a funny shape now and a bit frayed, but it does the job and keeps the sun off my head. A linen scarf stops my neck from burning.

It's cooler down here than being on the streets, though. There is a gentle breeze and the sound of lapping waves makes me feel less boiled than I am. A small group of pigeons is also cooling off in the shallows under the shade of Blackfriars Bridge, and the sip and jingle of a charm of goldfinches, known to the Anglo-Saxons as *thisteltuige* – 'thistle-tweaker', drifts down from a clump of wispy trees on the walkway. Like skylarks, whose eyes were put out in the cruelly mistaken belief that they sang better when they were blind, these brightly coloured birds were once trapped in the fields around London and kept in tiny cages for their beautiful song. So many were captured that by the end of the nineteenth century the number of wild goldfinches was seriously low. Thankfully they are a common sight now, even in central London, where they are finally free of their cages. The wings of the ones above me make a dry, papery sound as they buzz from tree to tree and I wonder if at least one has returned to the nest, a small tangle of twigs and moss lodged in the crook of a branch, that I saw back in December when the trees were bare.

On the foreshore there is a lot of plastic – the usual bottles, lighters, bottle tops and zip ties, along with a Bic biro, plastic straws, an asthma inhaler and two footballs. It's the season for lost balls again, now that people are outside. There is also a small Union Jack flag caught up in the shingle from last month's Jubilee celebrations. I pick up the balls and put them by the stairs for anyone who wants to take them and start to collect plastic. I try to take at least three pieces away with me, more if I can carry it, but it's an endless task.

I walk along the strandline slowly, stuffing pieces of plastic into my cotton tote bag. It is almost full when I step over a dark patch of mud and a tiny rainbow catches my eye. At first, I think it is another blob from the glassworks, but it is a lighter aqua-green colour and bobbled like a raspberry, which is what gives it its name: 'raspberry prunt'. It is sick and iridescent with age, but would once have embellished the side or stem of a glass called a rummer, which were made in Germany and Holland and peaked in fashion in the seventeenth century. Prunts were applied as molten glass drops and impressed with a prunt stamp to make the bobbly pattern. Drinking glasses were valuable, so as well as being decorative, the prunts provided better grip in the days before forks were commonly used when fingers could be greasy and slippery at mealtimes.

I pocket the prunt, tie the tote bag shut and clatter across a patch of flint cobbles towards the same stairs I'd come down earlier, pausing briefly to search a patch of mud that is bristling with old animal bones. There is one bone that doesn't fit. It's obvious. It's been worked and is a flatter, thinner strip than the bones around it. It is half buried in the mud and I gently pull it free. I wipe as much of the mud away as I can and peer through the grey smear it leaves behind. I can see a pattern of three circles of concentric rings with dots in the middle that are spaced out down the centre, and a carved line following the edge.

The pattern is known as 'ring and dot' and is probably the most common simple decoration found on objects dating from the post-medieval period to as far back as the Bronze Age, possibly even earlier. Any hidden meaning to the design has been lost to the mists of time, or maybe there never was any meaning. Perhaps it was just a comfortable pattern that appealed to a basic human desire for symmetry and regularity in an unpredictable world.

The rings and dots would have been made with a hand or bow drill using a toothed bit with a longer central spike to make the dot. The piece of bone is thin, and there is a small iron rivet in one end and a hole drilled in the other that suggests it is a scale from a knife handle, probably medieval. The bone is unusually dark, which may be staining from the mud or could even be dye. During the medieval period, bone was sometimes dyed black with a mixture of ferrous sulphate and oak galls, small round growths that form around the larvae of gall wasps.

The bone handle is delicate, so I wrap it in a small plastic ziplock bag to protect it and stash it safely in the bag around my waist. Then I climb slowly back up the wide concrete stairs into the sunshine and the heat. I pause at the top and look across the river to the foreshore at Queenhithe, where eight figures are sizzling in the sun, then I set off to the Millennium Bridge, emptying the plastic from my tote bag into a rubbish bin as I pass.

The tide has been rising for forty minutes and there is a row of silent people along the waterline, following it slowly back up the foreshore. They are seasoned mudlarks and they understand the unwritten rules. Beyond a brief exchange of pleasantries, we wait in silence, each of us focused on our own thoughts. Their heads are down and their backs are bent, but I see some familiar faces among them. I'm not wedded to one particular stretch, and I move around more than some mudlarks do, but it's not unusual for people to return to the same area again and again, convinced that if they look elsewhere they will miss something in their usual patch. They stick with their beloved spot through feast and famine. Some narrow this down even further and only ever search the same small area of a specific patch, where every dip and pebble becomes familiar, friendly and safe.

I find a quiet place away from the others to settle down. It's frowned on to kneel right next to someone else and conversation is never expected. If it does happen, it is usually kept fairly short, because tide and time wait for no man (or woman). The few people I do exchange words with complain about the tide and how it hasn't dropped as low as it should. A good low tide is below 0.5 m and today it is predicted to be 0.58 m, which is fairly good, but I can already tell it's going to be disappointing. Even something as predictable as the tide can be unpredictable sometimes.

Then Sean comes up to me, smiling. He looks hot in waterproof trousers and wellington boots, and he isn't happy with the tide either. 'Nothing today,' he says, 'but I did get Dave out of the mud at last.' 'Dave' is a huge seashell he'd first seen about fourteen months ago. 'It seemed to be fused to the wall,' he tells me, 'so I used to cover it up if I saw it. I called him Dave, then over the last month or so the area really started to erode and more of him appeared.'

Sean decided it was time to rescue Dave and asked Steve, a member of the Society of Thames Mudlarks with permission to dig in the area, if he would dig him up. As he lifted Dave clear of the mud, they saw the shell had been sitting on a Victorian silver sixpence.

The shell is almost a foot wide, and Sean has identified it as a *Turbo marmoratus* or turban shell. 'It needs cleaning up and years of mud scooping out from inside, but isn't it fascinating? I mean, how did it get there?' These snails live in the warm waters of the Indian Ocean, off Tanzania and Madagascar, the tropical western Pacific Ocean and off the coast of Queensland, so it is a long way from home. Scattered pieces of shining abalone shell can be found on this part of the foreshore that probably came from the shell warehouse that was beside the river in the nineteenth century. Perhaps Dave had come from there too. Turban shells were traditionally a source of mother-of-pearl for buttons, jewellery and furniture inlay, but before the nineteenth century they were rare curios that were turned into ornamental cups mounted in precious metals and much sought-after by wealthy gentlemen for their cabinets of wonder.

Sean mops his brow and walks away, but five minutes later he is walking back towards me with a grin. 'My day just got better,' he says, holding out a small silver coin on his palm. 'It just washed in on a wave at my feet – do you recognise it?' I do, and I covet it. I can see the twin shields of England and Ireland, which tells me it is a Commonwealth penny minted between 1649 and 1660, around the time of the English Civil War, when Britain briefly became a republic. Having removed Charles I's head in 1649, Oliver Cromwell, who ruled as Lord Protector of the Commonwealth from

1653 to 1658, also removed the monarch's head from the nation's coinage. I've yet to find a Commonwealth coin and I gaze at Sean's lucky find with green eyes.

I sweat it out for another hour, just in case the returning tide throws up another Commonwealth penny, but it doesn't. I find a George I (r. 1714–27) halfpenny instead, dated 1717, and a fifty-cent Euro coin. My final find of the day is a small shard of blue-and-white eighteenth-century Chinese porcelain with a hand-painted flower, possibly a chrysanthemum, a symbol of longevity and wealth.

I say goodbye to Sean and as I make for the stairs, I meet a man with a handful of lead type. It's a fairly common find on this part of the foreshore, and the jury's out on whether it washed down the drains from the print rooms of Fleet Street, London's newspaper district, tossed into the river by print workers clearing out their pockets on their way home after a shift, or found its way into the river in some other way. The man is a songwriter and he tells me he is going to look for hidden words among the letters. 'If there are any,' he says, 'I'll write a song about them.' If the river was to send a message through the medium of lost type, I think as I slowly climb the steep stairs under the Millennium Bridge, I wonder what it would say?

Tuesday 21 June 2022 (low tide 0.57 m @ London Bridge, 01.06)
Central London – North and South Banks

The longest day is bittersweet. It marks the start of the astronomical summer but also its demise. From now on the days will draw in and light will vanish as we begin our slow descent back to darker nights and colder weather. It is often thought to be an entire day, but it is in fact an exact moment, when the hemisphere is most tilted towards the sun. This year the sun will rise at 4.43 a.m., the solstice is at 10.13 a.m. and sunset is at 9.22 p.m. We will have sixteen hours and forty-three minutes of daylight. The night before is just seven hours and seventeen minutes long.

When I mudlark at night, I usually leave the foreshore before it gets too late and travel home on the last train with the drunk, depressed and merry. When I mudlark early, I rarely get to the foreshore before 5 a.m., so there are three to four hours that I have never spent with the river. Tonight, the tide is perfectly timed to lark through these unfamiliar hours and the added magical draw of being there on solstice eve has finally tempted me to do it.

I prepare nervously, pack and repack, research where to park, check the tide multiple times and, of course, the precise time the sun will appear in the sky. I'm larking into a new day for the first time and am not sure what to expect. Will the river be the comfortable friend I know from more sociable hours, or is there a darker side to it that only shows itself in the dead of night? Because I usually leave before the wee hours I have never really felt completely alone, but tonight I am larking into the quiet zone while the city sleeps.

I catch a couple of hours of sleep in the afternoon, before the kids come home from school, and get to central London at around midnight. The main roads are still busy, but the smaller roads behind the Tate Modern are quiet and empty. I park the car and walk quickly towards the river. It isn't far, but I am keen to get out of the gloom and away from the silent streets. A fox patters across the road in front of me, stops, looks me square in the face, then jogs over to a layby filled with hire bikes and vanishes. I am in his world now, an empty city filled with shadows and the day's rubbish that's skittering over lonely patches of grass and empty pavements.

I pick up the pace. I want to get off the deserted streets and onto the foreshore where I know I'll feel safe and hidden, protected by the Thames. I had planned to cross over the Millennium Bridge, but I can hear people approaching in the distance, talking loudly, probably drunk, and I make a snap decision to start at Bankside instead. I walk quickly to the railings, slide the catch back silently, slip through the low, heavy metal gate at the top of the stairs, close it as quietly as I can, and vanish like the fox.

I am a ghost now, one of many that haunt the foreshore at night: the drowned miller, the sailor far from home, the lovesick teenager, the starving mudlark. The veil between our two worlds is thinning. Some say it is at its thinnest on Midsummer's Eve in four days' time, when the spirits walk freely. In the past, people wore garlands of flowers at midsummer to ward off evil, but I don't have any to protect me. I don't sense anything bad, though, just a foreshore filled with the past that doesn't frighten me.

I stay in the shadow of the river wall with my head torch switched off until I reach the place I want to start searching. The foreshore at Bankside is more exposed and visible from the path above than it is on the north side, and there is enough city light to see where I am going without drawing attention to myself. I get to the patch I want, kneel down, flick on my torch and sweep the beam over the mud. My eyes are used to ranging freely in the daylight, but the torch restricts them to a much smaller area, which helps focus my search but is also frustrating. I see my first find quite quickly though, a large pin that I push into the lapel of my jacket for luck and maybe, in the absence of flowers, a bit of protection.

A man screams in the distance – I hope it is just someone playing around – and a party boat across the river turns up the music for one last song. I press on with my torch switched off, under the bridges at Blackfriars where the graffiti is so fresh I can still smell the paint. The lights that illuminate the bridges at night cycle from blue through green and orange to pink and back again, and my footsteps on the gravel echo, making me turn nervously to see if someone is following me.

I almost tread on my next find. It is a little bottle, around 2 in long, with a wide lip, a short neck and tell-tale ripples in the glass that reflect in the light of my head torch and tell me it is old blown glass. It must have just eroded out of the mud, which is probably what has preserved it, and it is lying on its side, half buried. A few more tides and it would have been plucked out by the river and broken or washed away altogether. I wonder if my sudden change of mind was the river drawing me down to rescue it.

I carefully ease the end of one finger into the mouth of the small bottle and gently pull, praying it is complete and not, as has happened so many other times, an illusion, a half-complete dream. It comes away from its muddy packaging with a sudden slurp and a sandy crunch, leaving behind a perfect impression in the smooth grey mud that oozes and slowly fills up with water. The spirits of the solstice are with me – the bottle is perfect. Even if I find nothing else at all, my effort has been worth it.

I walk down to the river to wash off some of the mud, and turn the bottle slowly in the light of my head torch. I've found scores of similar broken lips and bottoms, but this is the earliest complete free-blown apothecary bottle I have ever found. Its flared lip is wonky from being hand finished and there are no seam marks to suggest it was blown in a mould. The little scar on the base confirms that it was blown from a blob of molten glass on the end of a blowpipe and transferred to an iron bar called a punty or pontil while it was shaped and finished. When it was finally broken off the pontil, a scar was left, like a little glass belly button.

It is an eighteenth-century bottle and I expect the glass to be aqua green or clear, but it looks black, which is unusual. I wash it some more and look carefully. The black seems to be a coating inside the bottle, and I catch a familiar pungent smell coming from it. It is the thick aromatic scent of tar, probably pine tar, which was used to treat ship wood and rigging. It was also used to treat foot infections in horses and painted onto the wounds of pigs to encourage healing and cover the smell of blood, because other pigs will bite if they smell blood.

Tar is antibacterial and antifungal and has been used for centuries as a cure for skin complaints. The *Pharmacopoeia Londinensis* of 1788 includes a simple ointment made of tar that in a slightly modified form was still being listed in medical books as a treatment for ringworm, eczema and ulcers in the late nineteenth century:

UNGUENTUM PICIS.

Tar-Ointment.

Take of Tar,

Mutton-fuet [suet], prepared, of each
half a pound.

Melt them together, and ftrain [strain].

I don't find much else on the south side, but I feel much bolder by the time I am ready to head north over the Millennium Bridge. It is 1.27 a.m. and the river is still falling, thick and still, liquid black like ink. I turn off my torch and head towards the river stairs. The sky is a black bowl above my head, but the lights from the City are just enough to see by and to make out two dark shadows on the foreshore. I am not alone.

My stomach lurches. I tiptoe off the noisy gravel onto the sand, which is quieter, and creep closer. Two men are sitting side by side, sharing a can of beer; their bikes are laid down on the shingle behind them. We lock eyes in the gloom, I see fear in theirs and I suddenly realise why. A strange woman in a dirty jacket, wellies and knee pads has suddenly emerged from the shadows when they assumed they were perfectly alone. We stare at each other for a moment in silence, then I fix my eyes on the ground and head intently for the stairs.

Without street lights and away from the lights of the bridges, the foreshore on the north side is far darker. The velvet blackness deepens as I near Queenhithe Dock, where buildings edge the river and block out the City's light. This few hundred feet fizzes with the past in daylight, but at this silent hour the darkness is a whirligig of time, an all-consuming swirling soup of the past, thick, deep and loud, and for the first time the river and the foreshore frighten me.

I deliberate about moving further into the darkness. I want to get to the tunnel that runs under Vintner's Hall and out to the other side, where the foreshore opens up and there are lights from Southwark Bridge, but I am too scared. Instead, I dither on the edge of the darkness, searching the same patch over and over again and humming to fill the silence. When I stop humming, all I can hear in the darkness is the sound of the foreshore itself, which is always noisier at night. It is a strange electrical buzzing and popping, a sighing radio from deep within the mud. Ancient gases are escaping, history is rising, and the mud is communicating in its own morse code.

The light changes just after 3 a.m. It is almost imperceptible, a slight shift from black to the deepest blue; any satellites will have faded but I can still see stars and as I look up, something white flashes past. Gulls are circling silently above me, sharp white shapes like folded paper, origami birds that swoop and whirl in the gathering light. I take the lightening sky as my cue to finally move on. Inside the tunnel it is pitch black and the dripping water that's hardly noticeable by day is unbearably loud. I can't rush it, though, I have to move slowly: the sloping concrete is slippery and I can't risk falling in the dark, on my own with the tide coming in. I edge onwards into the gloom, my eyes stretched wide and my ears alert to every sound. Finally, I jump off the concrete and onto a pile of old roof tiles that clatter and echo back into the tunnel. I look around briefly and move quickly away from the tunnel's dark maw, into the lights below the bridge.

The new day climbs quickly into the sky, switching off the stars one by one. Within twenty minutes, indigo has turned sapphire blue, cloudless with just a single aeroplane vapour trail scratched into the sky, following the line of the river below. The word 'dawn' derives from the Old English verb *dagian*, 'to become day'. Twilight arrives in stages and is defined by the sun's position relative to the horizon. First comes astronomical twilight, when the sun is between 18° and 12° below the horizon and the stars begin to vanish. Nautical twilight, so called because the horizon is not visible enough for sailors to navigate by at sea, is when the sun is between 12° and 6° below the horizon. Finally, civil twilight, when there is enough natural light to see by, is from 6° below the horizon to sunrise.

In the past it was thought the hour between 3 a.m. and 4 a.m. was the time when this world and the next were closest. As if the peace that descends before dawn makes it easier to step over the threshold, to let go or give up. I lived with a doctor once who would tell me about her night shift over breakfast, and how this hour would steal the most souls – the death hour. She said it had something to do with blood pressure and hormones, but I preferred the notion of thresholds and welcoming hands.

On cue, around 3.30 a.m., my blood sugar drops, and I start to feel a bit wobbly. I don't want to risk fainting for the same reason I didn't want to risk falling in the tunnel, so I reach into my rucksack with a shaking hand, pull out a KitKat, sit down on a damp chunk of concrete and wait for the sugar to work. Then I wend my way under Cannon Street Bridge and by the time I am close to London Bridge it is light enough to mudlark without my head torch. I sit again in the new flat grey light and pour myself a cup of coffee from my flask. Beyond Tower Bridge, in the distance, the sky on the horizon is turning orange and the city around me is starting to wake up. I hear a door slam, a bin being wheeled out for the morning rubbish collection, and I see the occasional lonely figure crossing London Bridge.

I had thought I would welcome the dawn from the foreshore, but the bridge is blocking my view, so I retrace my steps to the stairs beside Cannon Street Bridge and climb wearily off the foreshore. A second set of steps takes me up to London Bridge and I walk to the very centre of it. It is so quiet that each time an empty bus passes by, I can hear the bridge creak under its weight. Two photographers are already there, waiting patiently by their cameras with their lenses trained on the horizon.

Below us, the river is a mirror reflecting the bridges and riverside buildings. It is so calm it looks still, but when I lean out over the wall and look down, I see a curious phenomenon. Clouds of silt are billowing up and blooming on the surface, curling then sinking again. I've never seen the river like this before and it is mesmerising, then something small and black breaks the mirrored surface. It moves and grows larger as it gets closer and I see that it is the head of a seal, a riverine selkie, a shape-shifter. It is believed they can cast off their seal skin to become human, and in Shetland they lure people into the water on Midsummer's Eve.

The seal turns lazily, flicks its tail and vanishes again. For the next hour I watch her play and the foreshore vanish. The photographers try to catch her in their lenses as the sun nears the horizon and the sky beyond the towers of Canary Wharf turns from orange to yellow. There is no great celebration on the bridge when the sun finally appears; the photographers merely adjust their settings. I leave it rising with the river and head for the car, feeling a little jetlagged and bewitched by the solstice.

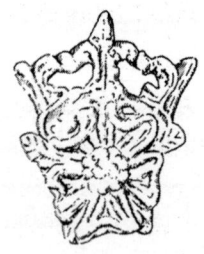

CODE: 127.22.G14

OBJECT:	15th century livery badge
MATERIAL:	Pewter
DATE FOUND:	12/07/2022
LOCATION:	Greenwich
NOTES:	A cast pewter badge, missing the pin from the back and featuring a crowned rose. It dates from the time of the Wars of the Roses and is likely to be a badge declaring affiliation to the Lancastrians, who were associated with Greenwich Palace. No sign of colour, but it is possible the rose was once painted red. Surface find.

JULY

Central London – South Bank

One day we will move back to our little two up, two down near the river, but in the meantime, I make the most of being able to return as cleaner, decorator and general fixer-upper between tenants. I'm in Greenwich this month to redecorate the house while Sarah holds the fort at home. With the river only five minutes away, I can visit the foreshore as often as I like, but today I have decided to hop on the train to London Bridge, just ten minutes up the track, for a few hours on the foreshore in central London.

The forecast is predicting a heatwave, and it is already getting hot. I mumble my pardons as I pick my way between a gaggle of teenagers sitting on the steps in the sun and stand for a moment in the cool shade of the river wall. The foreshore in front of the Tate Modern is busy with curious visitors and further along, where I was planning to search, it is scattered with mudlarks, many of whom I recognise as regulars from across the other side of the river.

I was hoping for some peace and quiet, but there are conversations all around me and it's hard to block them out. People are complaining they haven't found much, and I am not holding out much hope myself. The area has already been well scraped and searched, but every pair

171

of eyes sees differently, and I tell myself there is always a chance something has been missed.

I kneel down by the water's edge where the shingle has parted and exposed a fresh patch of small metal pieces, mainly nails, bolts, pieces of rusty wire, flakes and anonymous globs of rust, but all I find today is a single aglet, a brass tube that would have been sewn or crimped onto the end of a lace, much like the plastic ones that are attached to shoelaces today.

I crawl slowly on my hands and knees from the water's edge over shingle, mud and sand to the river wall, where I search another patch of small iron bits. Nothing. I walk back down to the river, pausing to pick up the neck of an eighteenth-century wine bottle with the cork still in place. Then I take a few steps to the west, kneel down and begin my slow progress to the wall again, executing stretches and poses that wouldn't be out of place on a yoga mat. I think about giving up, but I decide on one last sweep before I change location. On my third slow shuffle up the foreshore towards the river wall, as I peer around a large, smooth boulder, I see a flash of bright yellow in the mud. I can't believe they missed it. Gold!

I've been tricked by the foreshore before and pounced on plastic gold coins and the foil wrappings from chocolate coins, but real gold is unmistakable, especially really old gold, which is pure, rich and buttery. Even after centuries in the mud it emerges as bright and shiny as the day it was dropped. Most of the gold I've found is tiny and broken. Little shavings lost by goldsmiths that washed down drains, broken loops from chains, and snapped pins from brooches that fell into the mud unnoticed by their wearers. I found a gold pen nib and, once, a sixteenth-century gold-filigree aglet that was crushed and broken, but no less beautiful for it.

There is no denying that the glint of gold in mud is an exquisite sight and I've often wondered about our human obsession with this rare, star-brought metal. It has been hunted and coveted since prehistory. Gold nuggets were picked from the streams and hillsides of Wales, Scotland and the south-west and north-east of England. The Celts acquired gold from mines in the Mediterranean through trade, and the Romans brought gold from Spain and as far away as Africa. With the Age of Discovery, from the sixteenth century, gold was brought to the British Isles from all over the world, including captured bullion from treasure ships sailing across the Atlantic from the Americas to Spain. The most recent rush of gold, which was minted into millions of sovereigns, came from the nineteenth-century goldfields of Australia, Canada, America and South Africa, but still there is less of it than we might imagine. According to the World Gold Council, all the gold ever found could fit into a cube around 73 ft long on each side. What we have, has been melted down and reused over and over since prehistory. So, the gold chain around your neck may once have been a nugget in a Welsh stream, a Celtic coin, a Roman brooch, a medieval ring, a Georgian watch chain or a Victorian sovereign.

I sit back on my heels and look down at the tiny piece of treasure. I want to savour the moment before I touch it and I also want to be absolutely sure it is gold before I get my hopes up. It is a small, round, slightly domed disc, no larger than 0.2 in across, with delicate twisted decoration around the edge. It winks at me in the sunshine, and I think of all the eyes that had travelled over it that morning and not seen it. I am just the lucky one – there is no more or less to it than that. I looked in the right place at the right time when the sun was at the perfect height to give away its hiding place.

There is another mudlark close by and I call him over before I pick it up, I want to share my discovery. We kneel together and gaze in silence, then he sighs softly and goes back to his own patch, affording me another moment to myself. Only a mudlark would understand the need for privacy at a time like this. I bend down until my nose is as close to the mud as I dare and each grain of sand is a rock, every flake of rust an iron girder and the almost infinitesimal drop of water balanced on the edge of the gold is a lake. I know people around me are still chatting loudly, but all I can hear is silence.

It looks like a tiny button with a thin wire loop or shank on the back. If it is a button, it had to be for something very delicate, like lace or silk, but I have seen something similar. It was found on the other side of the river not so long ago and has been described and dated as a 'sixteenth-century decoration for clothing'. The Tudors, both men and women, were partial to a bit of bling, and covered their clothes and accessories with small shiny things. My piece of treasure was likely to be a kind of sequin, but in the sixteenth century people didn't use the word sequin.

According to the Oxford English Dictionary, the first recorded use of 'sequin' as a name for a decorative metal disc was in 1882. In the sixteenth, seventeenth and eighteenth centuries they were known as spangles or spangs, and were made in copper, silver and gold in a variety of shapes and sizes, although round was the most common. Oes were metal eyelets tacked or clinched to the material and often paired with spangles.

Spangles were sewn onto dresses and jerkins, also hats and shoes. In 1577 Elizabeth I was given a set of ruffs made of 'cut-work flourished with golde and silver and set with spangills'. Just three years earlier she had set her name to the *Proclamation against Excess of Apparel* to reinforce the laws that governed what people were permitted to wear, based on their social rank and wealth. Spangles were specifically mentioned as a decoration for the upper classes only: 'Cowls, sleeves, partlets, and linings, trimmed with spangles or pearls of gold, silver, or pearl; cowls of gold or silver, or of silk mixed with gold or silver; except the degrees and persons above mentioned; and trimmed with pearl, none under the degree of baroness or like degrees.'

My mind slithers through time. I see the little gold disc sewn onto deep red silk that rustles and sways as the wearer steps gingerly down a set of algae-covered wooden steps to a waiting wherry. She stumbles, the wherryman puts out his hand and the spangle, its thread already loose, catches on his sleeve. There is an imperceptible 'plip' as it drops into the water and flutters down into the mud.

Tuesday 5 July (low tide 1.19 m @ London Bridge, 12.26)
Central London – North Bank

A red admiral butterfly flits into my train carriage at London Bridge this morning and hitches a lift one stop to Cannon Street. It flutters out behind a line of hot, pink-faced commuters and takes off towards the sunshine at the end of the platform. I've bunked off the painting again today and I'm heading north. After the melee at Bankside yesterday, I have decided not to second-guess the crowds and to try my luck on the foreshore on the north side instead.

London is dry and dusty. Without rain, the weeds and moss at the top of the river wall have turned yellow and crispy, but my buddleia is flowering bravely. Three thin drooping spikes of tiny purple flowers hang limply from its virtually leafless, brittle branches. It looks exhausted and helpless, but it is not alone. All over London, hot dry wastelands and forgotten corners are filled with these wilting flowers and their heady scent of honey.

There is a dump of old red bricks on this part of the foreshore that I always search carefully on my hands and knees. Small objects like coins and buttons often catch among the bricks as the river swirls and eddies around them. The bricks themselves are handmade, there are marks on them from the wooden moulds that formed them, and the black mud they are embedded in is studded with the usual mix of old bones and oyster shells, but I search it specifically for the lead cloth seals that occasionally work their way up from between the bricks. The little lead seals, with numbers, initials, coats of arms and sometimes dates stamped into them, were once attached to bolts of woollen cloth to prove they were of quality and that the necessary alnage tax (a specific tax on woollen cloth) had been paid.

I have found hundreds of well-preserved lead cloth seals and almost all of them have come from this patch or the wider area around it. My seals date from the sixteenth to the eighteenth century, but most of them are from the seventeenth century. One theory for why there are so many here is that there were dyers working near the river, ripping off the seals before they plunged the cloth into the huge vats of dye made from indigo, madder, saffron and woad that produced fantastical colours such as 'pease-porrige tawnie', 'popongaie blue' and 'gooseturd greene'. The dyers also attached their own lead seals and I have a few of those too, but the really special cloth seals are those with an impression of the warp and weft of cloth woven centuries ago impressed into the soft lead on the reverse.

I am in luck today and add two more seals to my collection. The thin edge of one is just showing between two of the bricks. It is made of four small discs and dates from the early eighteenth century. On one disc there is a seated lion and the numerals 1½, which meant 1½d alnage tax had been paid on the cloth it was attached to. The other seal is older, probably late sixteenth century, and less well preserved. I can see an initial, W, and a *, which were probably part of the mark used by the merchant who had traded it.

Nobody else arrives and it's peaceful and quiet, so I stay and search on my knees with my fingertips for a couple of hours, then I'm ready to leave. I don't need to stay longer; I can come back tomorrow. I have this luxury and I'm going to make the most of it.

Wednesday 6 July (low tide 1.09 m @ London Bridge, 13.24)
Central London – South Bank

I get up early and paint the kitchen before I leave. I decide to risk a return to Bankside, hoping it has regained its peace and solitude… and it has. The area that had been so busy on Monday is completely deserted and I let out a silent cheer. There are no footsteps either, which means I am the first on the tide to search. Only the crows eye me beadily. Even two young mallard drakes sitting high up on the foreshore ignore me, with their beaks nestled tightly under their wings. Their green feathers are as iridescent as sick glass, but they still have fluffy patches of down on their heads and I wonder if they are the survivors of the brood I'd seen in April. Time is passing and spring's babies are growing up.

The temperature has dropped too, which is a relief, but there is a slight chill to the wind and I find myself shivering in my thin shirt as I crest the top of a low hill of rubble.

There is always coal here, sometimes very large lumps that swim up through the mud and glisten darkly in the sunshine. Mudlarks ignore them these days, but they were once highly sought-after by Victorian mudlarks, who could wash and sell them on the streets for a few valuable pennies. John Rocque's map of 1746 shows two coal yards on Bankside that fuelled the pottery kilns, glass houses, metal works, blacksmiths and, of course, the giant brewery. It also fired a huge riverside steam-powered flour mill called Albion Mills that burned down in 1791, much to the delight of London's millers, who blamed the mill for stealing their livelihoods and danced on Blackfriars Bridge in the glow of the fire wearing placards that read '*Success to the Mills of Albion but no Albion Mills*'. Its blackened remains are said by many to have been the 'Dark Satanic Mills' in William Blake's poem of 1804, 'And did those feet in ancient time'.

By the twentieth century, tons of coal were being shovelled into the great furnaces of the power station beside the river that is now the Tate Modern art gallery, but coal has been used as a fuel in London since the thirteenth century. It came from the north-east in seagoing boats, then by barge for its journey up the Thames to London. It was mainly used in lime kilns and by blacksmiths, but between 1570 and 1600 people began to use coal in their houses too. A surge in London's population at the end of the sixteenth century saw the price of wood rocket, so, ever resourceful, Londoners turned to dirty, smelly, smoky coal for their homes. The smoke it produced was as unpopular in the seventeenth century as it was in the nineteenth and twentieth centuries, with diarist John Evelyn describing it in 1661 as a 'fuliginous and filthy vapour'. London became a city of chimneys to channel the acrid, sooty smoke upwards and settled under a blanket of coal smoke that barely lifted until the Clean Air Act of 1956.

I work my way along the foreshore to the eroding bargebed, where I spot the unmistakable dome of a small emerging skull. Although most of it is still buried, I can tell it is bone from its warm honey colour and that it is likely to be a skull from its smooth shape and the tiny sutures that run over the top of it. I'd read about babies being abandoned in dust heaps and on the streets of eighteenth-century London by desperate women who didn't want them and couldn't afford them. It stood to reason that these small unwanted lives also ended up in the river, so I hold my breath as I gently clear away the mud and heave a sigh of relief when I reveal two eye orbits far too large to be human.

The skull is also the wrong shape. It is elongated, almost egg-like, and had definitely belonged to an animal. At first, I think it might have been a little lap dog, like the ones that became fashionable in the eighteenth century, but it is too small and the teeth aren't right, so I conclude it must be that of a cat. A mudlark once found a monkey skull on the river, a poor creature that was probably brought back by a sailor and died from the cold or from being fed the wrong food. I've found plenty of dog and cat jaws, but this is my first complete cat skull.

I bag the delicate skull still filled with mud to keep its tiny bones in place. I will clean it up when I get home and dry it out. The best place for it will be on my bookshelf, next to the dried frog I found a few summers ago, a shrivelled baby grass snake found on a woodland walk with the twins, two bird skulls that were washed up on our local beach and a bottle of dried bumble bees, collected from the garden.

I still remember the first dry toad I found. It had been squashed by a car and was stuck to the road outside my grandparent's house. I was allowed, in fact encouraged by my mother, to peel it off the tarmac and take it home. It seemed the normal thing to do and still does. In general, my family never questions the oddities I bring home. Only occasionally will Sarah ask if I really need to keep something that doesn't smell quite right or question where it's going in the house.

Thursday 7 July (low tide 1.0 m @ Chelsea Bridge, 14.35)
Westminster

I've been listening to the radio as I paint. The government is rebelling and Boris Johnson is being pressured to resign. I set myself a goal to finish painting the walls of the front bedroom by 10 a.m. so I have time to get to the foreshore at Westminster and to perhaps soak up something of the atmosphere.

It takes longer to get to Westminster from Greenwich than it takes to get to the foreshore in central London, but by 11.00 I'm making my way from the Tube to the river, walking behind the Houses of Parliament and past Westminster Cathedral. It's always busy here, but today it is chaos. As well as the usual crowds of tourists, there are urgent-looking people in suits carrying stacks of papers, news crews, lines of police, and demonstrators calling for the prime minister's resignation. A man is walking up and down shouting something through a homemade megaphone: a plastic four-pint milk bottle with the bottom cut off. It's not a very good megaphone because I can't hear what he's saying.

I turn off the road, away from the circus around Parliament and into the comparable peace of Victoria Tower Gardens. The gardens were created in the 1870s as part of Joseph Bazalgette's work for the Metropolitan Board of Works to provide London with a modern sewage system. Before it was built, there were warehouses and wharfs that went right up to Westminster Palace and boats and barges beached in lines on a long stretch of foreshore in front of them. The embankment ended this casual affair. It raised the level of the land and tamed the river, keeping it back with a tall granite wall. The old warehouses were pulled down and the land was reclaimed. Deep beneath the soil I am standing on, under the infill and the sewage pipes, there are river treasures that will never be found.

I walk to the wide, solid river wall and peer over the top. I have misjudged the tide. The water is low, but it is still tickling the bottom of the wall and there isn't any foreshore showing yet. I reckon I have at least half an hour to wait until I can get searching, so I take my time wandering through the gardens towards Lambeth Bridge and settle myself in a crook in the wall, shaded by tall lime trees, to watch the tide fall.

The stairs at Westminster are called Thorney Stairs, named after Thorney Island, an inhospitable eyot of shingle, marsh and thorns upon which Benedictine monks founded Westminster Abbey, around 960 CE. Palaces were also built here, the first by the Danish king Cnut (Canute) (r. 1016–35), who is said by some to have vainly thought he could control the river's tide with his kingly powers. Cnut's palace burned down and traces of the medieval palaces that followed were built into the Houses of Parliament, which was completed in 1870.

While I wait, I check the PLA's maps on my phone. They tell you where you can and can't mudlark, and the stretch in front of the Palace of Westminster is marked in black. For obvious security reasons, it is a Permanent Exclusion Zone and nobody is allowed to mudlark there. Other mudlarks have told me there are infrared rays that alert the police to anyone straying too close and a dog walker is rumoured to have accidentally set them off, but the problem is knowing where the foreshore becomes a no-go zone.

As the tide falls slowly, I look out across the water. The river at Westminster is shallow, only 2.1 m in places, which has led some to speculate that it is the point where Caesar's troops waded across in 54 BCE, towards the waiting Britons on the north shore. Nobody went further to prove this theory right than the eccentric experimental archaeologist

Lord Noel-Buxton, who in 1952 waded into the water at low tide wearing a knitted jumper, shirt and slacks, hoping to find a submerged path. Unfortunately, he didn't allow for the deepening of the channel over two millennia and only reached the second pier of Westminster Bridge before having to swim the rest of it.

When the foreshore finally starts to show at the bottom of the river wall, I climb a short flight of concrete steps to a low metal gate with spikes on top and an old sign that reads:

GREATER LONDON COUNCIL
WARNING
CHILDREN MUST NOT PLAY ON THESE STAIRS

The catch on the gate is broken and the hinges have dropped, so it won't open, but it is low enough to climb over onto the slippery steps down to the foreshore and I take them slowly. I tiptoe along a thin strip of gravel until I reach a wider patch of the foreshore just in front of Lambeth Bridge, where I start looking.

Fewer Clippers come this far, and they aren't as regular, so there is less erosion but there is the usual array of bricks, some broken corrugated asbestos roofing, lots of oyster shells, roof tiles, a broken drainage pipe, a few animal bones, and shreds of leather that were probably once old shoes. There are far fewer clay pipe stems than in central London and almost no large pieces of iron or nails. Most of the pottery I find is nineteenth century, but I do find some earlier pieces of eighteenth-century Chinese-import porcelain, delft and Wedgwood. The tide has left behind two strandlines of more modern detritus in which I find a bright red acrylic nail, and a pair of aviator sunglasses, too scratched to see through, otherwise I would have taken them to wear. I've found some of my best sunglasses in the river, presumably dropped off the faces of people on boats or peering over bridges and the river wall.

I walk on under the bridge and now I can see the hasty repair that was made to the river wall in 1941, when a high-explosive bomb from a night-time Luftwaffe air raid blew a hole in it. A temporary dam was quickly constructed by the Thames Flood Prevention Emergency Repair Unit to prevent what would have been a catastrophic flood. The Office of Works then set to repairing the wall properly. There was no time to carefully cut granite blocks to match what was there, so instead they filled the hole with concrete and scored it to imitate the granite blocks around it. It is still obvious to anyone who knows it's there, along with the scattered blocks of masonry that are fanned out on the foreshore in front of it.

From among the blasted chunks of grey-speckled granite, I pick up a piece that looks different. It is rough, sandy, buff-coloured stone that has been carved into a 'V' shape, and I can see chisel marks on one side. I know pieces of masonry from the medieval palace have been found here and it looks like a piece of tracery, the stone framework that made up medieval windows. It is certainly worth taking home, and since it is only palm-sized, I slide it into my rucksack, where it clanks against my water bottle as I walk.

Conscious that I am being watched by security cameras, I stay a very conservative distance from the Houses of Parliament and settle down to search among the shingle. Apart from the helicopter circling over the chaos outside Parliament, it is peaceful, with only the rustle of lime trees to disturb me. It smells fresher than the foreshore in central London too, greener and less urban, and I am joined by a crow that flies down with a slice of pizza in her beak. She is followed by two youngsters, grown-up babies that gape, flap and screech for food. She tears up the pizza and feeds each one in turn, then gobbles the last of it down herself. I wonder, with a slight pang of guilt, how Sarah is getting on at home with the twins and decide to call her later.

My finds from the day are an eclectic mix. An eighteenth-century clay pipe, an iron ox shoe, a prehistoric flint blade, a very rusty Victorian padlock and a broken brass penny whistle. It is said Thames watermen whittled whistles from alderwood to blow when the Thames fogged over; each fipple (mouthpiece) was slightly different, each whistle unique. I don't know if that's true, but I do know that drums were beaten from the shore to guide the wherries safely across in fog.

The last pieces I pick up bring me swiftly back to the here and now: a yellow 'I ♥ UK Parliament' badge and a broken House of Commons side plate. The green line around the edge identifies it as House of Commons and not House of Lords, which is red. Green serge was first recorded being used on the seats in the House of Commons as early 1632 and as I look up at Lambeth Bridge I remember being told at school that it is painted red for the House of Lords, while Westminster Bridge is painted green for the House of Commons.

Sunday 10 July (low tide 1.0 m @ Chelsea Bridge, 17.51)
Vauxhall

I am standing on the platform of the Victoria line at Stockwell Tube station, willing the train to arrive. There is a distant rumble, and the train is preceded by a blast of hot, stale, rubbery wind that disturbs the haze of dust hanging in the air and scares away the dusty black Tube mice that are looking for scraps of food under the tracks. It does nothing to ease the heat, though. The doors slide open with a clatter and more damp hot air from inside the carriage tumbles out with the passengers. Even though it is Sunday, it is busy, and I hold my breath as I squeeze in and find a space. The door hisses shut, we are sealed in and swept off into the boiling tunnels beneath London.

Even the polluted air over the Wandsworth Road is a welcome relief when I finally emerge from the subway. I am late to meet Flora and it is too hot to run, so I walk a bit faster and text her my apologies instead. Since I first went to Vauxhall years ago, the riverside has filled with tall buildings and hutch-like apartments piled one on top of another. It's a futuristic jumble of glass and steel fighting for space beside the river, a far call from the fetid alleyways, decrepit houses, wharfs, boat yards and stinking factories that tumbled towards the river 200 years ago.

Flora is waiting for me on a bench and pulling on her wellies. 'We can get over here,' she says, pointing to the wide granite wall. I look over it at a set of short iron rods, turned up at the end, that have been hammered in either side of a wooden bumper post, designed to prevent boats and barges from damaging the wall.

Flora has a PhD in Anglo-Saxon history and literature and an interest in just about everything. I don't usually mudlark with other people, in fact I try not to. They distract and sometimes irritate me, but Flora is easy to mudlark with because she is quiet and easy company. We usually bump into each other by chance on the foreshore in central London, but today we have met with a purpose at Vauxhall. I'm hoping to be able to show her the Saxon fish trap that is only visible if the tide falls low enough. To the uninformed eye it doesn't look like much – a long 'V'-shaped collection of wooden stakes sticking out of the foreshore – but once it's been explained, it's easy to imagine the wattle panels that went between them, and the fish being funnelled to a waiting net at the end.

The 'ladder' down to the foreshore is ingenious and only about twelve feet high, so I feel quite safe as I shuffle off the top of the wall and scramble down onto the shingle. A large gaggle of Canada geese are eating hot green algae off the wall and their beaks are making loud plastic clattering sounds as they snap at it. I give them a wide berth.

The foreshore is as I remember it... sparse. We walk east to the cofferdam near Vauxhall Bridge, which is a new addition since my last visit. It makes it hard to see the statues on the bridge, eight twice life-sized bronze women, four on each side, representing industry. I have to crane my neck and peer around the wall of the cofferdam just to see Pottery,

Engineering, Architecture and Agriculture. A thick lagoon of creamy mud has pooled under the bridge, and I throw in a stone to check its consistency and depth. It lands with a silky plop and I decide not to risk walking through it to see the ladies of Fine Art, Science, Local Government and Education on the other side.

Instead, we turn and head west, passing a lonely metal detectorist who barely looks up at us. I wonder if he is having more luck than we are, then I spot a pleasingly shaped stone that looks out of place among the flint pebbles. It is completely flat on one side, domed on the other and not a type of stone that I recognise from the foreshore. It fits perfectly in my palm, and when I turn it over, I see the flat side has been worn so smooth it is almost shiny. 'What do you think?' I say to Flora, 'a grinding stone?' 'Maybe,' she says, not sounding completely convinced. If it is, it could be prehistoric or it might be much later, since grinding stones have been used for thousands of years to crush, grind and pound everything from corn to pigments and plants. It is the only interesting thing I have found, though, so I drop it into my bag and crunch on through the pebbles.

We pass a hump of hot, dry shingle that is skirted by a tideline of plastic and driftwood. The hump looks as if it stays clear of all but the highest tides and a heat haze shimmers above it. At the top, growing from a wooden stump by the river wall, is a single stunted alder tree, a mysterious damp-dweller with timber that was prized for its resistance to rot in water. The bottom half is covered in algae and the small trunk that emerges from it is twisted and gnarled, but the river is keeping it alive, and its leaves are fresh, abundant and healthy.

I reach the end of the foreshore where the fish trap will appear if it's going to show today. Flora is still quite a long way away, wandering slowly and searching intently, so I find a patch of shade by the river wall and watch a crowded party boat thump past. Even in the shade it's hot, a kind of muggy stifling heat that's rising from the mud. I pull my water bottle out of my rucksack to take a drink, but the water is warm, and I start to feel suffocated and desperate to escape the heat. I need to wait for Flora, so I breathe the warm algae-saturated air slowly, in and out, focusing on her progress to take my mind off my rising panic.

By the time she gets to me with a handful of fossils I am cooler and calmer, and as I stand up to look at what she's found I glance over her shoulder towards the river. Two short black stumps are emerging from the water and the past is inching its way into the present again. 'Saxon fish trap!' I yell.

Monday 11 July (low tide 1.19 m @ North Woolwich, 05.55)
Greenwich

It is too hot to sleep, so I decide to catch the early tide before the city starts to heat up. I pull on a pair of shorts and a T-shirt, grab my boots from the back door and by 5 a.m. I am walking through the deserted streets of Greenwich towards the river.

I pass a bakery that is already pumping out delicious smells. The early morning light is like fire, reflecting off the glass dome over the entrance to the foot tunnel, and without the usual crowd of busy feet the pavement looks dirty and sticky. Pigeons are pecking through yesterday's rubbish, which is overflowing from every bin, and when I finally reach it, the river feels clean and fresh.

I stand for a moment looking down at it from the river path. A crew of rowers are on their way back to their clubhouse. They are just three black dots in the distance, rowing against the tide, but it is so quiet I can hear their conversation and the repetitive voice of the cox keeping time. The sun is rising, burning off a faint mist, and there isn't a breath of wind. The river is so calm and flat that an old barge moored just offshore appears to hover above the still water.

The stillness is beautiful, but it isn't good for searching. The Clippers have yet to start up and wash away the usual thin layer of sludge that covers the foreshore. Birds have left trails of footprints through the glazed gloop that shows it is quite thick in places, but I decide to go down anyway and take the muddy steps slowly, wishing I'd worn wellies instead of boots.

I've been watching the river at Greenwich for over twenty years and in that time I've seen it change and vanish. The base plates of the ancient jetty have washed away and the posts that are left are getting longer every year as the mud around them disappears. Soon they will wash away too. I felt mixed emotions when the PLA lowered huge nets of stones onto the foreshore in front of the wall, covering a fruitful patch, but it was essential to stop the river from undermining the wall. I had the same mixed emotions when they designated the westerly end of the foreshore a Scheduled Ancient Monument to protect it. It meant I could no longer search the area in any way, not even by eye, but it also meant that it was now protected.

Scheduling is a way of protecting heritage that began in 1913, although its roots go back to the 1882 Ancient Monuments Protection Act, when a 'Schedule' of mostly prehistoric monuments deserving state protection was drawn up. Greenwich isn't the only Scheduled part of the foreshore. Queenhithe Dock in central London, the oldest surviving dock on the Thames, was Scheduled in 1973, and the site of the launch of SS *Great Eastern* on the Isle of Dogs, Isambard Kingdom Brunel's most risky and final shipbuilding project, was Scheduled in 2015.

It is a criminal offence to damage or remove anything from a Scheduled Ancient Monument, but there will always be those who haven't bothered to check the maps, those who don't even know there is such a thing, and those who know but really don't care. There are no signs warning people about the protected areas and it's easy to drift into them so I sometimes draw a line in the sand with my boot along the edge of the Scheduled part of Greenwich to remind myself where to stop.

I paint all day in my hot and stuffy little house and return to the river for the evening tide and some fresh air. It is like the old days, when I would often sneak away from my desk for a couple of hours on the foreshore, twice a day if I could. As far as anyone else was concerned I was hard at work, but one day I was rumbled by one of Sarah's friends, who spotted me in the distance.

Text from friend to Sarah: 'Just seen Lara by the river. Looks so peaceful.'

Text from Sarah to me: 'Thought u said u were working today.'

Greenwich still feels like my part of the river and I'm very protective of it. I have poured hours of searching into the mud here, cried onto its shore and told it my deepest secrets. Coming back to it is like visiting an old friend: the shingle may have shifted, some of the old wood is gone but it is still the same friend, nonetheless.

I find more than I thought I would, an unusual pewter badge or mount in the shape of an ear of maize and four handmade dress pins. As I am leaving, I spot a large spade-shaped bone being washed in and out at the waterline. It looks like the scapula (shoulder bone) of an ox or horse, and in the middle, where the bone is as thin as cardboard, one perfect circle has been cut out and an attempt at another has broken the thin bone. It is a perfectly preserved, centuries-old moment of frustration.

The holes in the scapula are large enough to be from making button forms, plain bone discs with a small hole in the centre. They were used to make the cloth-and-thread-covered buttons that were popular in the seventeenth and eighteenth centuries, including 'death head' buttons that were characterised by a central X, like the crossed bones in a skull and crossbones. I have found nineteen bone and three wooden button blanks at Greenwich in various sizes, including some that would fit the holes in the scapula perfectly. I have always assumed they had fallen off jackets and waistcoats, but now I wonder if there was a workshop nearby making them. As I walk home across the parched lawns of the Old Naval College, I think about the workshop. Could it have been beneath my feet, or was it further inland? Could it even have been under my own house?

Tuesday 12 July (low tide 0.95 m @ North Woolwich, 19.29)
Greenwich

There is no sun today, just muggy damp heat and white clouds huddled so low I feel I can almost touch them. They seal in pressure like a giant lid. Dog days, the hot muggy days of summer that the Romans associated with Sirius, the dog star, which is high in the sky at this time of year. It needs to rain, but there is no rain forecast. Even my little house, whose walls retain the cold whatever the season, is so hot I had to move my blow-up mattress downstairs last night for some relief.

The evening tide is helpfully timed, but after a sweltering day of sanding and painting I am exhausted. I also have blisters from stupidly wearing my boots with no socks yesterday and I know that even the short walk to the river will be painful. But I need to get out of the house, so I summon the energy, find some plasters and hobble through the back streets on hot, sticky melted tarmac to the foreshore. I am glad I do. The river wall is still gently radiating the day's heat, but the sound of waves lapping on shingle is cooling and there is a slight breeze blowing downstream that hasn't reached inland.

In all my years of searching at Greenwich I have never found a coin older than 1963, a chunky twelve-sided brass thrupenny bit (3d piece). I've yearned to find the face of one of the Tudor monarchs who lived in the lost palace, but I never have. I have found dateable objects, though. The smallest clay pipes I have ever found were from Greenwich.

They date from around 1580 and are barely larger than the end of my little finger.

It's said Elizabeth I tried smoking, or what they called 'tobacco drinking', when it was first brought back from the New World. If she did, it would have been from a tiny pipe like the ones I found. She may also have eaten food made in the many broken bowls and cooking pots I've picked up. Her maid may have worn the shoe that was attached to the sole I pulled from the mud, and combed her hair with the broken boxwood comb that was lying beneath it. A royal prince may have played with the ceramic whistle, shaped like a mythical beast with long horns and ears, that came all the way from Flanders. His old tutor may have used the delicate ear spoon to clean out his waxy ears and the monk tasked with his religious education may have counted his prayers of penance and devotion with the ivory rosary bead I found washed up among ancient bones from many royal feasts. Of course, there is no way of knowing any of this. It is all pure imagination and conjecture, one of the delights of mudlarking.

My first dateable objects today are two sixteenth-century shoe soles, one left and one right. I can date them by their shape. The first is an outer sole, small enough to be from a child's shoe, with a narrow, worn heel and a wide, rounded toe, but the second is even more interesting. It is the inner sole of an adult's cow-mouth shoe, the type of flat, wide, square-toed pump that Henry VIII was often depicted wearing. Because it is an inner sole, the leather is thinner and softer and has taken on the impression of the original owner's toes and heel. It isn't uncommon to see these ghostly shadows on old shoe soles, but this one has an impression of the shoe's fabric lining. I can see the weave of woollen cloth, which would have made the shoe much warmer in winter months. It is even more unmistakable where there would have been more weight and pressure from the heel and the ball of the foot. It is a magical connection to a 500-year-old fellow Greenwich resident.

I carry on searching along the edge of the stone nets and see what looks like a small, flat piece of scrap lead. It looks like a 'shape', rather than a random piece, so, adhering to my mantra 'always pick it up and turn it over', I pick it out of the mud and take it to the edge of the river to wash it off, savouring the cool water on my hands.

What I see makes my heart leap higher than seeing gold in the mud. Although the surface is crusty with oxide, I can make out a crown on top of a rose. I open my green plastic finds box (an old fishing-tackle box) with shaking fingers and put it in a compartment that's padded with museum-grade foam. I snap the lid closed carefully and carry on searching, but it's hard to concentrate, knowing it's there. My mind is in my finds bag and all I want to do now is get home and look at it properly.

Back at the house, I use a toothpick I find in a kitchen drawer to clean it up. The crowned rose is clear, and I am sure it is a pewter badge of some kind, but I want to show Colin to confirm it, so I email him a photograph and in the time it takes for me to make a cup of tea, he has replied. 'It's a fifteenth-century secular badge, made of pewter in the same way that pilgrim badges were made, but as a symbol of allegiance rather than religion.' The Metropolitan Museum in New York has one in their collection that they describe as a 'Tudor badge', while the Museum of London describe theirs as a Yorkist rose. It doesn't look like a double-petalled Tudor rose to me, which means it has to be either Yorkist or Lancastrian and had probably been worn in the final decades of the Wars of the Roses.

The Wars of the Roses always confused me at school. I knew the battles had been fought between two branches of the same family for the kingship of the country, but beyond that I didn't know much, and up to this point I didn't really care. Now I need to reacquaint myself with what I failed to learn and to try and find a Lancastrian or Yorkist connection to Greenwich. I begin by writing down the names of all the kings, their dates, their queens and battles. Next to it I write down everything I can find out about the history of the palace at Greenwich during that time. Then I marry up the two with c. 1450, the rough date Colin said the badge had probably been made.

What I discover is exciting. The palace was a hotspot for Lancastrians prior to 1486, when the Lancastrian King Henry VII married Elizabeth of York, united the houses of York and Lancaster and founded the Tudor dynasty. The Lancastrian connection to the Palace was Henry VI (r. 1422–61 and 1470–71), who was married to Margaret of Anjou, the White Queen. She was queen consort to a pious, reclusive and often mentally unwell husband and became a leading force in the Lancastrian camp, determined to obtain the crown for her son Prince Edward. According to a book I have on the history of the palace, in 1447, Margaret and Henry acquired Bella Court, then a manor house beside the river at Greenwich, from Humphrey, the Duke of Gloucester (the king's uncle). They renamed it the Palace of Placentia and set about making improvements, including a pier so that boats could be boarded at low tide.

There are wooden posts on the foreshore at Greenwich that appear as the tide falls like a row of rotten teeth. Dendrochronology (tree-ring dating) has so far failed to date them, but they have variously been said to be Roman, medieval, Tudor and even modern. I wonder, though, if they are the remains of the pier that Margaret and Henry built, from which Margaret was rowed upriver to London, away from her useless husband, to continue her scheming? Had the badge fallen off one of her supporters as they handed her into a boat or from a simple servant girl, as she flirted with her beau beside the river, who was hoping her cheap little show of affiliation would afford her some protection in uncertain times?

Wednesday 13 July (low tide 0.88 m @ North Woolwich, 08.00)
Greenwich

I see a thirsty fox slink down the stairs this morning. He pads silently down to the river's edge, his russet coat burning in the early sun and his tail so low it brushes the mud and shingle. I freeze and watch him. He is completely at ease in his world and prince of all he surveys. Then the gravel beneath my boot shifts ever so slightly and for the briefest moment his sharp eyes lock on mine. I see the fear of a wild thing mingled with the confidence of a streetwise urban creature, and he flees.

Tuesday 19 July (low tide 0.87 m @ London Bridge, 12.36)
Central London – South Bank

England is baking under a heatwave and temperatures peak today at 40° C. My house is an oven, creaking, warping and popping in the heat, the old timbers sighing, and the tiles snapping. Its tiny back garden is a burnt desert with no shade, so I decide to ignore the government's warning to stay at home, stick an icepack in the waistband of my jeans and head for what I reason will be the comparative cool of the river.

The train is a furnace on wheels, but there are only two stops to London Bridge, where I step out onto a platform shimmering in the heat. Outside the station, the city is empty and lethargic. The usual bustle has melted with the tarmac, and I join a thin line of people making their way slowly along the shady side of the street. I take my time, convinced there will be a cool breeze waiting for me at the river, but I am wrong. Although I am on the shady south side, the river's breath is a hot dragon-like gasp that surrounds me and makes me even hotter. By the time I reach the foreshore I am so hot I am almost tempted to pick up the sodden sun visor I see lying at the water's edge, but instead I pour water from my flask onto my linen scarf, drape it around my neck and squeeze myself into the shade of the river wall.

The mud is more pungent than usual and the algae on the river wall smells hot and alkaline. Clouds of tiny flies crawl into my eyes and up my nose as I work my way slowly into the shade of Blackfriars Bridge, where the bend in the river and the tall buildings collude to block out the sun. The river is concentrated and slick under the heat, with an unusual amount of seaweed floating in it. The seagulls have had the same idea as me. The shingle under the bridge is covered with little white bodies, settled down comfortably in the shade, and they scream in fury as they fly away from me. I wonder why they don't take to the water, which looks inviting even to me, but they just fly to a spot a little further along and settle back down again. London is floppy, the heat has sucked the energy out of everything and even the birds and the river are slow today.

Friday 22 July (low tide 1.35 m @ London Bridge, 14.23)
Central London – South Bank

A series of mighty thunderstorms have cleared the air and cooled things down. The pavements are cleaner and the dust has settled. Everyone seems relieved and easier. A man walking his dog along the foreshore stops for a chat and even the drunks who sometimes lurk in the shadows under the riverside walkway bid me a cheery hello. The decorating in Greenwich is finished and I hum and whistle as I lark, thinking about nothing much at all.

I head east for a change, towards London Bridge, and stop to inspect the two hollowed-out elm-trunk drains under Cannon Street Bridge. One of them is made up of three shorter pieces with wooden wedges rammed into the joins to hold them tightly together. Elm was prized for its resistance to rot and was used extensively in wet conditions, including boatbuilding and log water pipes that were still in use beneath the streets of London well into the twentieth century. Each year a few more inches of these pipes erode out of the mud and the bark skin peels away. There are currently more elm drains at Wapping and further upstream in front of the OXO Tower, but they will all eventually wash away, like the one on the north side near Cannon Street Bridge, which was taken on a tide several years ago.

There are vanishing mooring points down this way too. An old cartwheel that's lost its spokes and some roughly hewn stone blocks with heavy iron loops that appeared quite suddenly. One is small enough for me to turn over, and I find its owner's initials carved into the stone underneath. It's hard to say exactly how old these stone moorings are, but the style of the writing suggests this one at least is a good few hundred years old. They are vanishing as mysteriously as they appear, though, and I hope they are not being taken.

There are more abandoned moorings on other parts of the foreshore. I've seen plenty of cartwheels come and go; reused timbers from ships that sailed the globe and fought great sea battles; anchors of every size, some with one side removed to avoid damaging boat hulls in shallow water; and huge ugly lumps of concrete, both square and round, like giant bath plugs... one pull could drain the river completely, imagine that!

There is a large slab of wood on the Rotherhithe Peninsula that I first saw several years ago. It has a big metal ring and 'RBB 1861' carved into it in deep letters. It was a mooring point for R. B. Byass & Co, a bottle merchant located nearby. The yellow drums that provide modern moorings for ships and barges strain against the tide and bob on the surface like giant corks. They are either screwed into the riverbed or anchored by sinkers. All the mooring points that are still in use today have names; some relate to their location while the origin of others – Chain Rock and Scars Elbow – have been taken by time.

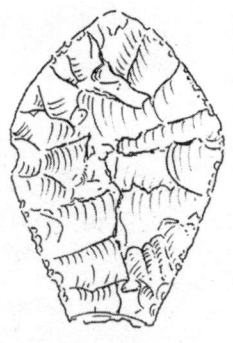

CODE: 68.22.Q23

OBJECT:	Neolithic 'leaf' arrowhead
MATERIAL:	Flint
DATE FOUND:	06/08/2022
LOCATION:	Central London – north bank
NOTES:	A leaf-shaped arrowhead of light brown flint. Very finely knapped in a process called pressure flaking. The wider end would have been hafted to the split end of a wooden arrow shaft. The pointed end is broken, perhaps on impact when it was fired. Surface find.

AUGUST

Thursday 4 August
The Society of Antiquaries Museum Room

The Society of Antiquaries of London closes annually in August, but I have managed to weasel a visit to the museum. I am meeting Kate, the society's curator, who told me that the collection includes some objects found in the Thames. I had seen the door to the museum on my first look around the society. It is large and heavy and has a wooden plate screwed onto it with the word 'MUSEUM' in gold letters. I decided there and then that I needed to see what was behind it.

The society's 'apartments' are in a nineteenth-century extension of Burlington House, a seventeenth-century mansion off Piccadilly in central London. The 'new' extension is home to the Linnean Society, the Geological Society, the Royal Astronomical Society and the Royal Society of Chemistry. The Royal Academy of Arts occupies the seventeenth-century mansion. By comparison, the entrance to the Society of Antiquaries is subtle and easy to miss. It is flanked by two potted box trees and watched over by the veiled face of a beautiful woman, known as the 'Veil of Time', which is carved into the keystone above the door.

I sign in and push through the heavy internal doors into a large, cool reception area. It has a faded, glued-together grandeur that is still impressive, and an elegant sweeping stone staircase that has also seen better days. Kate's office and the museum are on the third floor, so I begin to wind my way up, passing paintings of stern-looking eighteenth-century Fellows, a marble bust of George III, and a handsome longcase clock that ticks away the time it takes for me to reach the top floor.

Behind the large door, the room is smaller than I had expected and dominated by a wide, heavy wooden table. It smells as I had thought it would, though, of dust and books, paper and polish. The walls are painted a dark library red, and two sash windows are heavily lidded, with blinds pulled part-way down to keep out the light and heat. The bottom of one window is open a few inches and the blind is tapping against the window frame in the breeze. The only other sound is the occasional shuffle of someone walking past on the other side of the door.

My eye falls on the beautifully crafted Victorian rosewood cabinets underneath the windows that have rows of thin specimen drawers beneath low glass display cases. Opposite, on the other side of the room, are tall glass-doored cabinets with shelves that reach almost to the ceiling. I have cabinet envy. Both are filled with objects that I itch to look at, and I assume there is even more in the brown cardboard boxes piled up next to them. The room is stuffed, but there is a sense of order in this wonderful jumble of history.

As I look closer at the contents of the cabinets, I see some familiar faces. Three beardy Bellarmines, high up on a shelf, look back at me. I spot a green-glazed medieval jug, a seventeenth-century onion bottle, and I recognise the swirling pattern on a medieval costrel (a flat-bellied drinking flask) from shards I'd found on the foreshore.

There is a collection of pocket watches piled up next to a line of Roman oil lamps, Greek urns and an Egyptian figurine with crossed arms and a striped headdress. A carved stone gargoyle sits next to a cannon ball and above them there are three shelves of Roman pots that look just like the ones I've found on the Thames. There is another Roman pot on the mantelpiece above a large marble fireplace. Alongside are more boxes and bags of things and, tucked away behind the door, there is a small collector's cabinet, about two feet high.

'I think there's stuff in this room that hasn't been unpacked since we moved from Somerset House in 1874,' Kate says, seeing me looking at it all with wide eyes. Between 1780 and 1874 the society was based at Somerset House, overlooking the Thames on the Strand. When they moved, 45,000 objects went with them: 22,000 prints and drawings; 13,000 brass rubbings; 11,000 seal casts, original wax seals and matrices; and 2,000 archaeological objects.

'I worked at the British Museum for five years before I came here,' she says. 'Compared to the museum, it was quite daunting to begin with because the records are so sparse.' If an object was presented at an ordinary meeting, it was noted in the minutes. There are around fifty thick ledgers of these, all written in exquisite copperplate handwriting that looks beautiful but is hard for the modern eye to read. One of Kate's jobs is to go through these ledgers to try to find, identify and properly catalogue the objects that are mentioned. 'Even if we don't have a find spot, if we know the name of the donor we can find out where he was working or living and get an idea of where the object may have been found,' she says. 'But it's a huge task.'

She hands me a battered red cloth-bound book. 'We also have "the Red Book" – it is the most important record we have of the collection,' she says as I open it and read the title page: 'CATALOGUE OF ANTIQUITIES, COINS, PICTURES AND MISCELLANEOUS CURIOSITIES IN THE POSSESSION OF THE SOCIETY OF ANTIQUARIES 1847, COMPILED FROM THE MINUTES OF THE SOCIETY'S PROCEEDINGS, AND OTHER MEMORIALS, BY ALBERT WAY ESQ., F.S.A.' On the opposite page, written in pencil, is: 'Additions by Beatrice de Cardi, Fellow 1981–89'.

Albert Way's original book is a simple list of objects with some information taken from the minutes of meetings. Most were donations, but some had simply been left behind by Fellows who had brought them in to show and discuss. Beatrice de Cardi was the daughter of an American singer and heiress and a Corsican count. Her love of archaeology stemmed from visits to ancient sites in Corsica while on holiday, and she went on to become a leading figure in twentieth-century archaeology. She went back through Way's book in the 1980s, identified the objects and gave each one a code, which she painted directly onto them in minute numbers. Her handwritten and typed notes were pasted and inserted into Way's book. This was the first time that any of the objects in the museum had been properly catalogued since the society had been founded in 1707, but it was far from finished.

I try to code my own finds, though I'm far less organised than Beatrice. My codes are a combination of numbers and initials, and they correspond to my notebooks that log the find spot, date found, age of the object and any other known details. There is a level of secrecy to my notes and codes that I admit I revel in. I like being the only person who understands them, it is a record of my secret world, but I suppose I should be more transparent. In their current state I am the only person who knows what they mean, and I have a duty to the objects to hand them on to their next custodian with their journey properly recorded.

Kate had laid out some objects on the table. 'According to Way's book, most of these came from the Thames,' she says, 'but there is probably a lot more that hasn't been recorded.' There are Bronze Age swords, axe heads and spear heads, a medieval sword and two beautifully polished prehistoric stone axes. One of them has a paper label glued to one side that reads:

Dr Roots no.1
This beautiful specimen of the early British celt was
taken[?] at the Chelsea Water Works on the banks of the
Thames at Kingston in July 1855

Above it, in neat white letters, Beatrice had painted 'Cat 56'. It had been donated to the Society by a Dr Roots, who also bequeathed a ring of eight medieval keys that were found under a foundation stone of Old London Bridge when it was demolished in the early nineteenth century.

The only other thing on the table is a bag numbered 581. I open it and take out twenty-three smaller bags. Inside each is a medieval pewter badge. Their entry in the Red Book is a typed addition by Beatrice: '581. 23 pilgrim badges and 2 fragments, many from London. Late 13th–15th cent.'

'There's a good chance that some of the badges were found in the Thames, but there's no written record of it,' says Kate. I look at them carefully and remember what Colin had told me about the colour of Thames-found badges being quite distinctive and usually darker. They must be river badges, I think, as I run my fingers over a heraldic badge featuring a strange-looking squirrel eating an acorn. However perfect and beautiful they are though, without proper records all of them are still lost, their origins severed, and their histories cut painfully short.

I walk slowly around the edge of the room peering into cases, then, with a jolt of excitement, I spot three crude lead figures and a lead medallion that I recognise as the handiwork of Billy and Charley, fakers who had famously duped the Society 180 years ago. I'd seen some Billy and Charleys on display at the Museum of London, but it was even more special to see them here because there was a good chance they were the actual pieces that fooled the Society.

Mudlarks traditionally sold any antiquities they found on the foreshore to gentlemen collectors and antique dealers, and around the mid-nineteenth century the market was suddenly flooded with a large number of supposedly medieval lead objects. They were being offered for sale by two men, William Smith (Billy) and Charles Eaton (Charley), who claimed to have found them at Shadwell, where a new dock was being dug.

The objects were childish in style and poorly made, knights in armour, kings wearing strange spiky crowns, daggers, medallions, statuettes and ampullae. Where there were inscriptions, they were a nonsense jumble of letters and numbers because Billy and Charley were illiterate. Nothing quite like them had ever been seen before, but that didn't stop some 'experts' from proclaiming their authenticity, and all the while Billy and Charley were laughing up their sleeves at the establishment.

The boys were running quite an industry from their premises in east London. They cast the objects using plaster of Paris moulds that they carved with nails and knives. They finished them off in a bath of acid to simulate ageing, then passed them through a network of antique dealers who sold them to middle-class collectors, often people with more money than knowledge. They even succeeded in fooling some of the most eminent collectors of the time, and it was a good ruse. The materials needed to make a medallion like the one I was looking at in the cabinet cost just two pence, but it could be sold for half a crown (2s and 6d).

Not everyone was taken in, though, and the appearance of so many rare artefacts aroused suspicion. Secretary to the British Archaeological Association, Henry Syer Cuming, was extremely dubious, and his scepticism was shared by some of the keepers at the British Museum. In April 1858, in a lecture, Cuming condemned the objects as a 'Gross attempt at deception' and this was reported in the magazine *The Athenaeum*. Sales of Billy and Charley's handiwork crashed, and one disgruntled antique dealer called George Eastwood took action. He sued the magazine for libel, claiming they were accusing him of passing forgeries when he was selling them in good faith.

The case was held in August 1858 and Eastwood was first to testify. He claimed to have paid another antique dealer £296 for 1,100 lead objects, before buying more from Billy and Charley directly. Charley didn't appear in court, but Billy did, and was described by a reporter as 'a rough looking young man'. Billy claimed they had found 2,000 objects in total and made £400 (almost £41,000 in today's money) from their sale. He said they bribed dockworkers for them with drinks and searched for them themselves in the docks after hours. Expert witnesses, including Fellows of the Society of Antiquaries, archaeologists and antique dealers, were called to authenticate the objects, and the case of libel was rejected. A verdict of 'not guilty' was returned, since there was no evidence that George Eastwood had actually been alluded to in the article.

The debate about authenticity rumbled on for a while, then in 1861 Charles Reed, another Fellow of the Society of Antiquaries, exposed the fraud. A sewer hunter had offered to sell him some of the 'finds', and Reed had won his confidence, getting him to admit the objects were forgeries and introducing him to Billy and Charley. Reed bought their stock of artefacts and paid the sewer hunter to break into their workshop to steal the moulds. This was the final piece of evidence he needed to expose Billy and Charley as fraudsters. In March 1861, he exhibited everything to the Society of Antiquaries and the game was up... or was it?

Billy and Charley somehow eluded prosecution and continued to deceive and swindle with their forgeries. In 1867 they were arrested in Windsor after a clergyman recognised the objects they were selling and called the police, but there were insufficient grounds for prosecution and the pair fled back to London. By 1869, however, they were finding it so hard to sell their creations that it was said they could be bought for as little as a penny. Charley died of tuberculosis in a tenement in east London on 4 January

1870, aged thirty-five. Later that year, Billy failed to find a buyer for a badge bearing a picture of the Lamb of God that he later confessed to having copied from a butter mould. He vanished from history after 1871 and his fate is unknown.

I have great respect for these two enterprising opportunists. Most of the Fellows and Victorian gentlemen who had collected the contents of the boxes and cabinets in the museum room hadn't got their own hands dirty in the process. They had relied on ordinary people like Billy and Charley to do the hard work for them. While the names of the men who acquired the objects live on in the books and papers they wrote, the invisible army of finders, too poor to be intellectual, have mostly been forgotten: fishermen who dragged up ancient pots and bottles in their nets, dredgermen who found coins in the bottom of their buckets, farmers who revealed prehistoric flint tools and Roman statues with their ploughs, mudlarks and scavengers.

Kate pulls a few boxes off the shelves and leaves me to look through them. I start with the one labelled 'Prattinton Collection'. The information in the Red Book is basic, but I manage to find some of it listed. The objects were the eclectic gatherings of Dr Peter Prattinton (1771–1840), who is described as a gentleman antiquary. His collection is contained in tiny wooden boxes, wrapped in sheets of thick paper sealed with red sealing wax, and zipped into modern plastic zip bags. I work my way through them, discovering the wing bones of a swan, a sixteenth-century lock of hair from Abbot Lichfield of Evesham Abbey, human nasal bones, fossilised wood, a medieval iron spur and a 'portion of sacking' from a Knights Templar grave. It is the type of collection that speaks to me: random, varied and eccentric.

I pack Prattinton's treasures away carefully and look again at the little wooden collector's cabinet behind the door. 'What's in that?' I ask Kate when she comes back in to see how I am getting on. 'To be honest, I don't know,' she says. 'It was left to the society by a Fellow when he died. It's always been there – I'm not even sure if it's in the Red Book.' 'Can I look inside?' I ask. 'Sure,' she says, 'I think the drawers are quite stiff, though, you might need to wax them to get them back in,' and she vanishes again, returning with some furniture wax and a rag.

Kate is right. The thin drawers creak and squeak stiffly as I pull them out. Beatrice had been at work in the first two and numbered all the objects that were there. They tally with her notes in the Red Book, which reads: 'Wooden cabinet with door and 8 drawers containing an antiquarian collection of curiosities including a mourning ring, a brooch, strike-a-light, 18th cent knives and forks, flint implements, bronze weapons and objects from Cyprus, Majorca and India. List lodged in top drawer. Bequeathed by W. P. D. Stebbing, F.S.A 1961.' I look for the list, but it isn't there.

The little set of drawers is a cabinet of curiosities, delightfully known in German as a *Wunderkammer*, a 'room of wonder', from the early collections that could fill entire rooms. Wunderkammers became a thing in the sixteenth century, when men of means and intelligence amassed collections of rare and curious objects that they felt reflected themselves and how they saw the world. But there was often no rationale to them, and the contents of each room or cabinet varied enormously according to the collectors' interests.

Two types of collections could exist within one wunderkammer: *naturalia*, natural objects like shells, skeletons and rocks; and *arteficialia*, manmade artefacts, such as scientific tools, works of art and antiquities. Wunderkammers fell out of favour in the eighteenth century, as public museums began to open, but they gained popularity again in the nineteenth century, among the rising middle class who had the time, money and education to spend on indulging their curiosity.

My own wunderkammer as a child was an old chest of drawers that lived in the barn. In it, I kept my special 'treasures': dried toads, shells, animal skulls, bird's eggs, snake skins, interesting stones, fossils, a stag's antler, dried butterflies and broken Victorian crockery I found in the garden bed near the kitchen. These days I keep my finds in an eighteen-drawer oak printer's chest that was a lucky buy from a local junk shop. I've lined each drawer with a different coloured felt and divided up my river finds by type and age: a drawer for glass, one for Roman finds, pottery shards by date, bone artefacts, and another for beads and buttons. It sits in the corner of my office and every so often I'll spend an hour or an afternoon lost in its contents, reorganising or puzzling over mystery objects.

I work my way slowly through Stebbing's collection, feeling the same excitement and curiosity I did when I was allowed to rifle through my grandmother's sewing box. There is far more in the cabinet than was listed in the Red Book and the objects are wonderful and eclectic: a Bronze Age socketed axe head; glass phials of tiny molluscs collected from British beaches; a Roman spearhead; a small pot of gold dust; a chunk of amber; a cigarette holder; the ear bones of various fish; a string of ancient Egyptian scarab beetles; fossil shark teeth; a piece of sixteenth-century stained glass;

seventeenth-century clay pipes; some unusual locks dated 'Cyprus 1915'; and Dr Arthur Rowell's collection of old glass marbles that belonged to his 'grandmama who died in 1834'.

It dawns on me that I'm not just looking through a collection here, I am delving into the mind of a very curious man, and it feels quite intrusive, a bit like reading his diary. Collections are so much more than a gathering of things: they are physical representations of people's past, tastes, interests, memories, holidays and days out, connections and friendships. I know my own collections tell the story of me. They are a three-dimensional catalogue of where I've been and what has been important to me at different stages in my life.

By the time I reach the eighth drawer, the Society is preparing to close. I can hear people leaving the Fellows' Room, doors opening and closing and papers being shuffled. I have spent five hours in the company of past collectors and gatherers and barely scratched the surface. I leave with my head as stuffed as the little museum, full of other people's treasures and wondering about my own mudlarking collection. Where would it go? Who would want it? Would it end up in a museum, in a cardboard box on a high shelf, or in a wooden cabinet behind a door?

Friday 5 August
Collecting

I can't get Stebbing and his collection out of my mind. I need to know more about him, so I sit down at the computer and start to google. It doesn't take long to find a screenshot of his obituary in Vol. LXXVI of the *Archaeologia Cantiana*, a journal that has been published by the Kent Archaeological Society since 1858. William Pinckard Delane Stebbing had lived just a few miles along the coast from where I live today. Born in 1873, he was an engineer and an architect, with a passion for archaeology and antiquities, but his collection went way beyond the little wooden box on the floor of the society's museum. I also find him on the British Museum's website where, as well as a short biography, the 744 objects he bequeathed the museum are listed, from Celtic coins and seventeenth-century delft pots to Anglo-Saxon beads and medieval fishhooks.

I wonder where and how his collection began. All collections start with one object, but how many objects make up a collection? It has to be more than one, but is two enough? Could two objects with the intent to collect more be called a collection? And when is a collection complete? Most collectors would say never. There will always be gaps and tangents that keep them searching, because at the core of most collections, including my own, is anticipation and obsession, the thrill of the chase and the euphoria of finding the perfect piece.

I've always collected. Ever since I was very small, I've felt the need to assemble things that interest me into satisfyingly beautiful and comfortingly ordered groups. My mother still tells the story of the moss I gathered when I was about three. I chose each one very carefully and kept them in a little basket that I insisted on taking everywhere, even to the supermarket, and showing to anyone who would spare me the time.

The psychoanalyst and collector Werner Muensterberger suggested the impulse to collect begins in everyone as child, collecting objects like teddy bears and comfort blankets to substitute for their mother. I must be part of the estimated third of the population that continues collecting into adulthood. Muensterberger also said that collecting eases anxiety and uncertainty, which I suppose I agree with. It certainly makes me feel happy and relaxed, and filling the gaps in my collections gives me purpose and a goal.

Over the years, my collections have come and gone, been disbanded, given away, lost in house moves and stored in attics. I've revived old collections and started new ones, each one reflecting who I was at that point in time. When I travelled around south-east Asia in my early thirties, I decided to collect antique opium weights, setting myself a sensibly low price limit for what I was willing to pay. My search for them took me down backstreets and into the kind of local markets and shops I probably wouldn't have otherwise visited, which is another benefit of collecting. Most collectors will tell you it's more about the hunt than the quarry, and with that comes adventures and chance meetings. I came home with a small handful of opium weights that I subsequently lost in a relationship break-up, but I still think about them. Each one was hard found, carefully chosen and haggled over, each one has a story of which I am part.

Mudlarking is a very different way of collecting though. The things I find have never been part of another collection; they are fresh and new, unsullied by auctions and sales catalogues, and my hunt is indiscriminate, like a lucky dip. I take what the river offers, I don't search for anything specific. Consequently, my collecting is sporadic, and my collections grow slowly and organically. It has taken me twenty years to collect four pinner's bones and it might take

me several more years to add another, but I'd never buy
one or even accept one from another mudlark. That would
defeat the purpose of the collection. Everything needs to
have been found by me on the Thames foreshore, so that
I know all there is to know about where it came from and
the years between when it was lost by its original owner
and the moment I picked it up.

Gathering stuff in types helps me to create order in
chaos. Within my large and rambling foreshore collection,
I have more specific mini-collections that include (current
numbers): shoe soles (x 64); metal clasps and hinges
from sixteenth- and seventeenth-century books (x 29);
nineteenth-century glass bottle stoppers (x 94); lead cloth
seals (x 419); seventeenth-century clay pipes with maker's
stamps (x 32); eighteenth-century wig curlers (x 18); Roman
bone game counters (x 11); thimbles (x 17); human teeth
(x 4); medieval silver pennies (x 10); Bellarmine faces (x 33
complete, x 101 partial); and clay marbles (x 29).

I've got a bit of a thing for handles too. They are the
working bit of a pot, the most intimate part that was
touched and held by hands as it was passed around a table
or carried from the heat of a kitchen into a dining room.
My collection of around fifty-eight includes the broken
handles of jugs, cups, porringers, cooking pans, tankards
and bowls, and each one is different. The hollow tubes of
pipkins (three-legged cooking pots) and frying pans are
often blackened with soot on one side from the last fire
they cooked in. I have delicate porcelain teacup handles; the
flat wide straps of medieval jugs; chipped delft; thick, rough
red ware with thumbprints at the bottom where they were
pressed onto the pot; flat, latticed porringer handles; and a

strange yellow handle with frills around the edge that looks like a sea slug.

I don't collect indiscriminately, though, and I curate my collection carefully. I mostly collect objects that are hard to find, so my collection grows slowly, and I only keep an object if I don't already have one. If I have multiple versions, each will be slightly different. I also replace objects if I find better examples and I return the ones they oust to the foreshore, as close to where I originally found them as possible.

Because I take things back and leave so much behind, I have a virtual collection of the things I didn't take in my mind, and vivid memories of where and when I found the ones I took home. I can forget someone's name in a moment, but I clearly remember the cluster of marbles I found at Blackwall eight years ago and the large triangular wool weight I found in the gloaming of a winter's evening with my friend Helen.

Many mudlarks can tell you when and where they found their precious things, what was happening that day, how they got to the foreshore and who they showed it to first. They are not just objects: there are emotional investments and memories, and when the finder dies, part of the collection's story dies with them. Even if they pass it on to someone with the same interest, the collection will never be the same.

According to the Portable Antiquities Scheme, the issue of private collections is a time bomb. With metal detectors being so cheap and widely available and so many people detecting, there are more antiquities in private hands than there ever have been before, and some quite significant collections that risk losing context and valuable information when they are finally dispersed. The same is true of the collections built up by mudlarks in the golden years of the 1970s and 1980s. The generation that dug them up is getting old now and families aren't always interested in inheriting their finds.

Stuart was recently invited to look through the collection of a deceased mudlark. His family had got permission from the PLA to auction it, once the Museum of London had selected the more important archaeological artefacts that they thought would 'enhance their collection and add to their understanding of London and its inhabitants'. 'There were some incredible things there,' said Stuart when I spoke to him about it. 'Thankfully he was very organised and had labelled each object with the date and find spot, but there will be plenty of collections coming up in the next few decades that aren't so well organised.'

When Ivor Noël Hume died in 2017, I wondered what happened to his magnificent collection. Having read so many of his books, I felt a personal attachment to his Thames-found objects. In his lifetime he donated to the Chipstone Foundation, which was set up in the 1960s by two American collectors to preserve and interpret their own collection and to stimulate research in the decorative arts. I wondered how much had gone to them and if any was recorded as coming from the Thames.

Tina, the registrar at Chipstone, replied to my email: 'The Hume gift(s) started in 1999, but the majority came to us as they were handling his estate, with several shipments coming in 2018 and 2019.' She searched the current catalogue for me and came up with just seven objects listed as

Thames-found: a plain red earthenware fourteenth- to early fifteenth-century jug; three small, plain white seventeenth-century delft ointment pots; the base of a seventeenth-century delftware pot featuring the head of a cavalier painted in blue; two late sixteenth-century Bellarmine bottle necks with bearded faces; and a complete Bellarmine bottle with a record of its discovery. It was found near Blackfriars Bridge sometime between 1953 and 1955, and was retrieved by one of Hume's archaeological volunteers, known only as 'Johnny' Johnson, 'a strange, saturnine man reminiscent of one of the lost souls drawn by Gustave Doré in his studies of the destitute of Victorian London'. The bottle was cracked and should only have been worth a few shillings, but Hume paid more for it because it was a witch bottle, containing pins, a cloth heart and a wisp of brown human hair.

I knew Hume had hundreds of objects from the river, so where had the rest gone? I tracked some of them to an auction house in Asheville, North Carolina, but only a few lots looked as if they might have originated from the Thames. If I could have smelled them, perhaps I could have detected a faint whiff of river mud, but there was no mention of where they had been found in the catalogue. They had already lost their provenance, but one lot in particular caught my eye, an ornate eighteenth-century shoe buckle and two circular badges described as 'thin hardware'. They were Billy and Charley's.

I've always wanted a Billy and Charley because of the story, a true David and Goliath tale. My mudlarking collection isn't just a group of inanimate objects, it is a gathering of stories and Billy and Charley's is one that I would like to add. It seems Hume felt the same way and had managed to acquire two for his own collection, but these days buying a real Billy and Charley is risky. There are fakers making fakes of fakes now and they are very good. I wonder what the boys would have made of that, and I wish I'd known about the sale. Lot 122, sold on 5 December 2019 for just $100 – it was a bargain.

Saturday 6 August (low tide 1.19 m @ London Bridge, 1.38)
Central London – North and South Bank

I stagger to the foreshore with a big bag of stuff, a mixture of bones, shells, pipe bowls, pottery shards and miscellaneous things to give back to the river. I scatter them liberally and then head for the spot where I hope to find more of the 'Queenhithe Hoard'. For years, I have been collecting shards of eighteenth-century pottery from an area around twelve feet square. It appears in unpredictable waves; sometimes there's not much and other times the surface is littered with it. I don't know why it's there or where it came from, but I've recently discovered it's been washing up for decades.

In his book, *If These Pots Could Talk*, Hume describes a visit to the foreshore with his wife Audrey on a particularly low tide where they found a quantity of 'high-quality English porcelain', all dating from the last quarter of the eighteenth century. According to Hume, they were resting together in an area around two feet square, where he presumed they had been dropped or dumped. The accompanying image showed a group of thirty-two shards of types that are still found in this area today and he had helpfully worked out the percentages of each type, which roughly corresponded to my own observations over the years:

English porcelain	30.5%
Creamware	23%
White salt glaze	13%
Nottingham	10.5%
Pearlware	7%
Dry-bodied redware	7%
Engine-turned, glazed	3%
Debased scratch blue	3%
Chinese porcelain	3%

I walk slowly towards the other mudlarks who are already there. I know most of them; they are sharp-eyed and have probably picked up most of what is worth collecting, but I still see some of the hoard: Chinese and English porcelain, scratch blue, black basaltware and the broken lid of a Staffordshire redware tea or coffee pot with a pretty sprigged flower design.

Then I spot a light brown sliver, just larger than an old 50p piece. It is slightly convex, and the sun is reflecting off the irregular facets on its flat side. Incredibly, even though there are footprints all around it, everyone has missed it. It is a light brown leaf-shaped flint arrowhead.

I have found plenty of Mesolithic blades and scrapers before, but I have only ever read about these delicate and expertly made Neolithic arrowheads. They required great skill and were shaped with an antler or bone tine that removed pieces of flint, flake by flake, in a technique known as pressure-flaking. I hold it up to the clear blue sky; it is so thin the light shines through it and I can see scores of tiny chips across its surface where the flint has been flaked away.

Despite its fragility, it has survived for around 5,000 years and only the tip is broken. What had I ever made that would last that long?

Tuesday 9 August (low tide 1.64 m @ North Woolwich, 17.41)
Greenwich

We are gripped by drought. Statistics from the Met Office show that July was the driest in England since 1935, and the driest on record for the south and south-east of England. It hasn't affected water levels in the tidal Thames, but they are much lower in the non-tidal part above Teddington Lock. The natural spring that supplies the river, known as the source, and which dries up most summers, has dried up further downstream than ever before. The official start point of the Thames outside Cirencester in Gloucestershire has moved five miles further downstream.

There have been great droughts before. In 1540 there was an extreme heatwave across Europe and a mega-drought that lasted for eleven months. Wells and springs dried up, mills stood still and people and animals starved. In London, the Thames was so low that seawater extended above London Bridge, even at ebb tide. In Paris there was so little water in the Seine that people were able to walk across it, stepping around puddles filled with thick green water and bloated dead fish. Just think of the otherwise hidden treasures that were revealed.

I've thought quite a lot about the Thames drying up and imagined the swathes of artefacts. There would be the foundations of old London Bridge to explore, perhaps even the remains of the Roman bridge, and sunken ships, wherries and barges that might still be filled with cargo. I've thought about tiptoeing around bombs from World War II, scrambling over bicycles and cartwheels, dodging skulls and ribcages, and gawping at contraband: the guns, knives and other accessories to crime that have been thrown into the river over the centuries. I've even daydreamed about finding the Great Seal of England, the monarch's official seal that is said to have been flung into the Thames by the deposed King James II as he fled to France and exile in December 1688. But in reality, I suspect there would just be piles of rubbish, old iron, rubble, animal bones and coal to sift through. And mud, lots and lots of mud.

Friday 12 August (low tide 0.79 m @ London Bridge, 09.06)
Blackwall

Blackwall, directly opposite the O2 Arena, is a quiet and isolated stretch of foreshore that is virtually impossible to access. I know of only a handful of other mudlarks who have been there, but in all the times I've visited, I've never seen another soul. It is a mudlark's paradise, quiet, peaceful and virtually untouched, undug and, as far as I know, mostly undetected. I can mudlark there for hours, safe in the knowledge that I am unlikely to be disturbed.

For several years I got to Blackwall through the kindness of a local resident, who let me through a locked gate so that I could access a ladder, but that was the easy bit. The ladder stopped almost six feet short of the foreshore and the distance grew year by year as wash from the Clippers ate away at the foreshore beneath it. I devised an ingenious way of scaling it with a packable climbing ladder, but over lockdown I lost contact with my source, and I haven't been able to visit Blackwall for some time. Recently, though, I have found another way to get to this inaccessible part of the foreshore and today I am going back for the first time in a while.

I wake up early to catch the tide and leave a trail through dew on the lawn as I walk to the car. It is the first heavy dew of late summer and a sign that autumn is coming. I drive to Greenwich and park near the river. It is quicker to cycle to Blackwall from Greenwich, so I have brought my folding bike with me. I clip it together, pedal over to the Greenwich foot-tunnel entrance and take the lift down, revelling in the naughtiness of cycling all the way to the other side with no consequences.

I follow the river path east, around the long right arm of the Isle of Dogs, and in fifteen minutes I am at Blackwall. It is good to be back, and I pull on my gloves over the glove lines on my wrist. I have a mudlark's tan from a hot summer spent beside the river; my forearms down to my wrists and the back of my neck are golden brown, but the rest of me is distinctly paler.

At one end of the Blackwall foreshore is the Gun pub, once the popular haunt of dockworkers, stevedores, watermen and smugglers, who used the pub to offload their contraband before continuing upriver to the Port of London where taxes had to be paid. Today, the Gun is a gastropub for well-heeled bankers and office workers, and the foreshore in front of the pub's terrace is scattered with modern crockery and cutlery. Nothing has changed; the cutlery and crockery is still there, lying in a tangle of old rusted metal, tools, horseshoes, anchors and cables that were dumped into the river when it was an industrial area and home to workshops and boatyards.

The middle of the foreshore at Blackwall is dominated by the old River Police station, which closed in the 1970s and has been converted into apartments, like everything else. On the high tide the river washes underneath the building, depositing fine sand onto the old slipway where the police boats were once pulled up. The sand stretches down to the river in patches between the shingle, and as the tide recedes, it leaves jumbled strandlines of river-washed objects.

There are Victorian pottery shards aplenty – fresh blue-and-white patterns and thick earthy stoneware from broken bottles and jugs – and old dry bones that lie alongside fresh, greasy, creamy-white bones from the restaurant at the Gun. There are vulcanite bottle stoppers worn to shapeless nubs and among it all I find fifteen seventeenth-century clay pipes. Because nobody collects them, they roll around and wear away, eventually to nothing. All fifteen are well worn, smoothed and becoming shapeless, so I leave them where they are for the river to claim completely.

I weave my way down to the water's edge, the regular crunch of gravel a soothing chant from the foreshore, and almost step on a chunky handmade bone domino. Another day, another foreshore, another tide and I would have missed it. So much of mudlarking is about chance.

The domino is lying spots down and I pick it up. It has eight spots on one half and two on the other. I've only ever seen dominoes with up to six spots, but I've read there are larger sets running up to nine spots (fifty-eight pieces) and even twelve spots (ninety-one pieces). It is a simple, fairly unremarkable find, but something I've always wanted because of its history.

Dominoes originated in China and are mentioned as early as the tenth century. They are thought to have been brought to Europe by Italian missionaries and were introduced to England by French prisoners of war towards the end of the eighteenth century. The prisoners whittled sets of them from old bones and sold them for money or extra food. This little domino may have been such a piece, carved from an old soup bone on board a prison hulk and exchanged in a set for a loaf of bread with a passing sailor. It may have travelled the globe and been used in games played from Caribbean beaches to the bustling port of Shanghai, only to slip between the boards of a jetty at Blackwall.

I look out across the river at the O2 Arena and the jumble of buildings that are growing around it. When I moved to London in the 1990s it was a fairly desolate place, dominated by huge gas-storage tanks. Now it's a completely new residential area, or 'village', as the developer has branded it. I squint across the bend in the peninsula to a lost village, Orchard Place, which couldn't have been further from the opulence opposite me.

Orchard Place was one of the river's most isolated communities. It was a gathering of run-down terraced houses, pubs, a chapel and a school on a spit of land bounded on the south side by the Thames, on the north and east by Bow Creek, on the west by the East India Dock. The people who lived there worked in local factories and on ships, caught shrimps and scavenged the river from rowing boats for flotsam and jetsam they could sell. Their poverty and isolation led to a reputation for lawlessness and inbreeding – at one point, out of 160 children registered at the school, a hundred shared the surname Lammin. Even a local clergyman condemned them as 'incarnate mushrooms' and 'hardly human', but despite this, their community thrived for well over a hundred years, until the houses were condemned as slums in the 1930s and they were rehoused in nearby Poplar.

On the foreshore, in front of the houses that back onto the river at Blackwall, is a thick puddle of hardened tar that reaches from the river wall almost to the water at low tide. It is a twenty-foot amoeba that swallowed up broken pottery, stones, glass, wood and bricks as it eased its way down the foreshore. In the summer it softens and releases the ancient scent of sailing ships, and in the winter it hardens around its treasures. Anything prised from this great stinking fruitcake is useless, sticky, smelly and black, but below it, on lower tides like today, there is a line that yields scraps of lead, poorly finished musket balls and a mysterious number of broken and dismembered lead and pewter toys, some of which date back to the seventeenth century. There are also melted blobs of lead that set as they fell and cooled. Maybe someone was recycling lead here. Perhaps mudlarks and scavengers, or the residents of Orchard Place, took their paltry finds to one of the houses above the foreshore – pathetic, muddy sacks of scrap lead exchanged for a few grubby pennies from a greasy, long-whiskered man who haggled and bargained hard with them.

I walk up and down the waterline a few times, pull a Victorian shoe sole out of the mud, wash a curved shard of

Staffordshire slipware in a muddy puddle and pick up a lead bag seal. I find the wide bottom of an eighteenth-century wine bottle next to its broken neck, then I spot a coin in a patch of gloopy mud that is about to be overwhelmed by the wash of a passing Clipper. There is nothing to do but step off the chunk of concrete I am standing on and up to my ankles in gloop. It oozes down the top of my boots, I squelch three steps, snatch up the coin, and retreat quickly before I am submerged in water too. It will be a muddy cycle back to the car.

The coin turns out not to be a coin at all, it is a token with a hole through the middle and it is well worn, but at home I cover it with a thin sheet of paper and take a pencil rubbing that reveals enough detail to identify it. It is a seventeenth-century trade token from the White Horse Tavern in Greenwich that would have been handed over the counter by the landlord Hugh Pudefourd, who'd had his own coinage made with the words 'HIS HALFE PENNY' proudly announced in the centre.

The White Horse was tucked away down an alley off Fisher Lane that ran parallel with the Thames, behind waterfront buildings and close to the ruins of the old medieval palace. It features in Captain Marryat's novel *Poor Jack* (1840), where it is described as 'a very narrow street and what is said in a room on one side can be heard on the other ... there were drunken men and drunken women and occasionally scolding and fighting'.

The best thing about the token, though, is the nail hole through the centre, because it means it comes with a story: perhaps a sailor or bargee who nailed a token from his favourite tavern to the mast of his ship or barge as a reminder of home or for protection at sea.

My shadow creeps along the foreshore like a long, gangly alien. The light has changed, summer is slipping away and the city is tired and dusty. The streets are dirty and unwashed, the weeds between the cracks are burnt and even the wind is dry, but the sun is kinder and the sky is vivid blue. The searing heat of summer has gone. It has been replaced with a deep, golden heat that makes me want to stretch out and soak it up like a cat. The angle of the light at this time of year and the long shadows it casts are just right for spotting even the tiniest hidden things. These are perfect mudlarking days.

AUTUMN

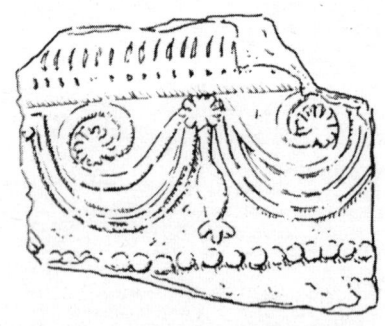

—— CODE: 139.22.SW18 ——

OBJECT:	Roman samian (*terra sigillata*)
MATERIAL:	Clay
DATE FOUND:	13/09/2022
LOCATION:	Kent
NOTES:	A shard of slip-coated, orange samian, made between the 1st and 3rd century CE. Finely decorated with a moulded floral 'festoon' ending in a 'bottle bud'. Probably part of a large, high-status bowl. Surface find.

September

Monday 5 September (low tide 1.89 m @ London Bridge, 14.38)
Central London – North Bank

Last night there was an almighty thunderstorm that finally broke the drought. I lay in bed listening to it rumbling towards us and counted the pauses between the blue flashes that lit the room and the sudden cracks that got louder and closer until we were in the eye of the storm, and it was shaking the leaves from the trees. Rain fell in a great deluge onto the parched lawn. It filled drains, overflowed gutters, and the land sighed as it drank, releasing smells of rehydrated soil, wet chalk and damp vegetation that crept into my bedroom through the open window.

There is an actual name for the smell that rain makes on dry earth. It is 'petrichor', a blend of the Greek words *petra* (stone) and *ichor* (the golden fluid that flowed through the veins of the ancient Greek gods). When a raindrop lands on dry, porous soil, small bubbles of air fizz to the surface of the water droplet and burst out, releasing aerosols. The scent contains oils secreted by plants during dry weather and chemicals released by bacteria that live in the soil. This is the essence of the earth and the smell of life.

The foreshore today is storm-washed. The river has lost the viscosity that had developed over the drought. Its thick, pungent, alkaline soupy smell has been replaced with the stench of wet sewage and it is filled with drought detritus. Larger objects that would otherwise have found their way gradually into the river have been swept up en masse and sent eastwards. A flotilla of logs, tree branches, plastic bottles and footballs bob past and a thin layer of bright-green duckweed, washed down from further upstream where standing water had overflowed into the river, clings to the shoreline.

The river's breath is foul. The smell tells me the storm drains opened last night and the tideline of cotton-bud sticks, dental cleaners, condoms and sanitary towels confirms it. When it rains, London's Victorian sewers, which were built to serve seven million fewer people than they do today, can't cope, and they overflow into the storm drains that lead directly into the river. The Walbrook, one of London's lost rivers that flow secretly under the city, is usually little more than a dribble through a drain in the wall near Cannon Street Bridge, but today it is running faster than I have ever seen it, and a foul grey gush that smells of sewage and stagnant rot is pouring around the iron hatch. I hate to see the river sullied like this; it brings me down and it makes me angry. My proud, dignified river is literally choking on our shit.

Wednesday 7 September (low tide 1.68 m @ London Bridge, 18.47)
Rotherhithe

I descend into the gloom beneath the Mayflower's outdoor terrace. Each time I come here, the drop from the river stairs onto the foreshore is longer and it won't be long before the landlord has to do something to make his outdoor deck safe. In places, erosion is creating cavities so large you can actually crawl into them. The Thames is always trying to escape, testing its boundaries on every tide and pushing its confines, biding its time. One day walls will collapse, maybe even whole riverside buildings, and the river will triumph.

The tide is just beginning to draw back over the higgledy-piggledy ship timbers that stick out of the mud at all angles. They are the remains of ships that were dismantled and recycled here centuries ago: wide planks and rudders, keel pieces, deck beams and thick, post-like windlasses with square holes to take the stout poles that were used to turn them. The river is picking the timbers out of its skin too. In the time I have been visiting, new pieces have emerged, while others look as if they could be lifted up and taken away by the tide at any moment. I often wonder how much is buried in the mud below my feet and how many more ships will surface and sail away with the river. Sometimes my mind continues down even further into the tunnel directly beneath me. While the men on the foreshore were tearing apart old ships, all kinds of debauchery was going on below in 'Hades Hotel'.

By the turn of the nineteenth century, this part of the river was a forest of masts and so densely packed with ships that it was said to be possible to walk from one side of the river to the other on their decks. The nearest bridge, London Bridge, was two miles away and often so congested that it could take two to three hours just to cross it. A new bridge at Rotherhithe would have prevented the tall-masted ships from sailing any further into London, so it was decided to dig a tunnel beneath the Thames instead.

The Thames Tunnel, which links Rotherhithe and Wapping, took eighteen years to dig by hand, through clay, gravel, quicksand and stinking river mud. The men tunnelled like moles behind a tunnelling shield that was divided into thirty-six cells, one per man, to prevent a total collapse. As they dug, confined to their own little worlds, they were followed by an army of bricklayers who shored it up behind them. They risked all four elements as they worked: ancient poisonous and flammable gases, roof collapses and floods that could fill the tunnel within minutes. Six men died while they were digging the tunnel and it's thought many more contracted cholera and dysentery from the river water that constantly showered them as they worked.

The project was overseen by master engineers Marc Brunel and his son Isambard, but it was plagued by financial difficulties and never became the major crossing they hoped it would be. By the time it opened in 1843 there wasn't even enough money left to build the large spiral roads that were needed for horses and carts to access it. Instead, it opened as a pedestrian tunnel, a shopping arcade, a place of entertainment and a visitor attraction, with 50,000 people, including Queen Victoria (r. 1837–1901), paying a penny to walk through it on the day it opened.

They were entertained by sword-swallowers and fire-eaters, magicians, female acrobats, Indian dancers, performing horses and Chinese singers, but within a few years the tunnel had earned a sinful reputation for criminality, poverty and vice, and the nickname 'Hades Hotel'. It became a haunt for pickpockets, prostitutes and gay men, and an underground drinking den where wild parties were thrown beneath bilious-yellow gas light. And among all of this, the city's destitute and homeless bedded down in the damp and darkness as a safer alternative to sleeping on the streets. In 1865 it was sold to the East London Railway to be used as a railway tunnel and then as a tunnel for Underground trains, becoming the oldest tunnel in the oldest underground system in the world.

I search for a while around the old ship timbers. It is muggy and suffocating, the sky is low, grey and foreboding, and there is a storm brewing in the dark clouds. I have a lingering headache behind my left eye, so it is something of a relief when the pressure finally lifts and there is a short, violent thunderstorm. Rain begins to fall in fat drops and I run to the Mayflower deck for shelter. The wind that comes with it blows away the pressure and blasts rain into my hiding place. It pours off the wood above me, splashes onto the foreshore and spritzes my face.

As the rain slows and the sun comes back out, I emerge from my hiding place and turn left, towards Bermondsey. By the eighteenth century the riverside along this stretch was lined with ramshackle buildings that virtually grew from the foreshore. Many were torn down to make way for wharfs and warehouses and of those that were left, few survived the Blitz during World War II. By the end of the war just a single row of eighteenth-century riverside dwellings was left standing, and by the 1950s they were run-down, crooked and dilapidated. Despite this, their character and river views attracted artists, writers, celebrities and even a princess.

In 1959 the photographer Antony Armstrong-Jones sublet a downstairs room in 59 Rotherhithe Street, the house at the end of the row. It was a simple, draughty, river-damp room with a magnificent view of the Thames, and when he began to court Princess Margaret, it became their sanctuary and hideaway. It was somewhere they could be ordinary people, cook together and drink cheap wine, away from the pomp and ceremony of her public life. But it didn't last long. In 1964 the houses were condemned by the council and demolished. Before they went, Margaret and Antony cleared out their love nest, taking everything but the mattress they had bought second hand for £1, which they threw into the river as they left.

There is a large patch of foreshore, close to where Princess Margaret's secret room used to be, that is very hard. I'm not sure if it was made that way or if it's just tons of iron nails and scrap that have bonded together over time. Whatever it is, it's filled with small metal things and pins that gradually break free and wash together in little prickly patches. Sometimes I find them still partly encased, tantalisingly close to freedom. I've found early lead toys, coins, buttons, cufflinks, beads and watch-winders here, most of which date from around the eighteenth and early nineteenth century. My reward for braving the storm today is a single late eighteenth-century copper alloy cufflink, hand-engraved with flowers.

Thursday 8 September
Collecting Finds From Jo the FLO

Jo the FLO is leaving for another job, so I nip up to Maidstone to collect a box of finds he's been recording for me. There are a few surprises and two objects of 'note', which is like getting a gold star for your efforts. He wants to hang on to the pot-shaped bead that he had thought was Anglo-Saxon, but now thinks is Late Iron Age to early Roman, and some of my medieval floor tiles, because he wants to see if he can identify the makers.

The objects he recorded were:

A counterfeit James I sixpence dated 1622 – a find of note that will be included in the *British Numismatic Journal* 'Coin Register'.

A Roman bone hinge with the wooden dowel still in place – another find of note that will be included in the next edition of *Britannia*, the foremost journal for the study of Roman Britain.

Two shards of Roman pottery.

A large shard of sixteenth- to seventeenth-century Spanish or Italian maiolica.

A post-medieval pinner's bone.

A large, triangular terracotta weight, either a Roman loom weight or a post-medieval net weight.

A Roman ceramic tuyère or bellows nozzle.

Two decorated medieval floor tiles.

A piece of fifteenth-century pilgrim badge featuring the Virgin Mary.

A piece of fifteenth-century pilgrim badge featuring John Shorne, the rector of North Marston in Buckinghamshire.

A fifteenth-century St Osmund of Salisbury pilgrim badge.

A seventeenth-century pewter toy dripping pan.

A late Iron Age or Roman bead (to be confirmed).

A shard of Roman Samian featuring a 'phallus dog'.

A whetstone, dated Roman to post-medieval.

A seventeenth-century pewter gunpowder-flask cap.

Three broken Roman bone hairpins.

A fifteenth-century knife handle.

A sixteenth-century cast copper-alloy button in the shape of an acorn.

A decorated sixteenth-century cast copper-alloy button.

A decorated sixteenth-century pewter button.

A counterfeit William III silver shilling.

I get back from visiting Jo and go straight out again to celebrate the twins' exam results. As we sit in the restaurant waiting for our main course to arrive, news of the Queen's death ripples through the room. There are gasps, hushed conversations and a woman on the table next to us begins to cry. I find myself crying too, which confuses me because it is irrational. I am not crying for a woman I never knew; I'm crying for the past, a fear of change, the loss of time and the passing of a generation. The moment she died she passed into history and became Elizabeth II (r. 1952–2022).

I saw her on 25 April 2012. I remember the date because it was the day the twins were born. They were supposed to be due on my birthday, but they arrived five weeks early. It was pouring with rain as we stood beside the river at Greenwich, waiting for the Queen to arrive. She came by boat, as Elizabeth I would have done when she visited her palace, and a fanfare of trumpets announced her arrival. We craned and stretched with everyone else and saw a tiny figure dressed in red in the distance, making her way through the rain to the old ship. The *Cutty Sark* was one of the fastest ships on earth in her day and had been dry-docked at Greenwich as a visitors' attraction since 1957. Over the intervening years the ship had suffered in her concrete prison. She rotted, corroded and sagged, cruelly trapped just yards from the water, where she had ruled the waves in the nineteenth century. A huge renovation project had returned her to her former glory, albeit still stranded and alone, and the Queen was there to reopen her.

We stayed until Sarah was too uncomfortable to stay any longer, then made our way home through the quiet, wet streets. It's funny the things you remember about momentous days. I can remember making a cup of tea to warm us up and asking Sarah if she wanted to go upstairs to put the cots together. At the top of the stairs her waters broke and much of the rest is a blur, but three hours later two tiny babies were born, and our lives changed for ever.

Tuesday 13 September, Fieldwalking
Kent

The word 'harvest' comes from the Old English word *hærfest*, meaning 'autumn'. It is the season for gathering and for fieldwalking, searching by eye on land rather than beside a river. I take time away from the river to fieldwalk in the autumn, after the harvest has been gathered and the soil is bare, before the new crop is drilled and there has been some rain to settle the soil. The recent storms have settled the earth without making it too muddy, which is perfect for showing up anything the weather, worms and plants have brought to the surface over the months since it was sown in April.

The field is close to a known Roman settlement, and I usually find at least one piece of Roman pottery when I come here. Last year the field was planted with wheat, and it was hard to find much between the stubble until the field was ploughed, but the farmer has told me this summer it was filled with neat wide rows of maize that have been scythed off at ankle height, leaving perfect rows of bare earth.

I get to the field mid-morning and without having to worry about the tide, I dither along the hedgerow for a bit, looking for autumnal treats. It is burnt and brown from the drought and the blackberries are dry and past their best, but I manage to find a small handful and five fawn-coloured hazelnuts with peachy-soft shells that miraculously survived the squirrels when they were green and tender in the summer.

I eat the blackberries as I begin to walk between the lines of maize stumps. My eyes are well trained for the foreshore, but fieldlarking needs a slight adjustment. There is less to see and analyse, dismiss and assess, so I can walk faster than I do when I'm mudlarking and cover more ground. I know I will find less, but that doesn't bother me, I'm outside on a beautiful day and who could ask for more?

There are plenty of orange lumps that stand out against the brown soil. Most of them are probably bricks and tiles of varying ages that have been broken up by the plough; there are also old flaky oyster shells that are always a good indication of human settlement and activity. For an hour and a half, I walk contentedly up and down the rows, absent-mindedly collecting rough grey Roman domestic ware and pieces of amphora. Behind the cottages at one end of the field I find broken glass twinkling in the sunlight, animal bones from Sunday lunches, and pottery, mostly Victorian blue and white, that I leave behind. The cottages are old and it's likely they tipped their waste into the field behind them.

In one stubbly furrow I find the bones and feathers of a racing pigeon, its identification ring still encircling a scaly leg. There are feathers scattered all over the field, but they aren't just from the dead pigeon. There are oily black crow feathers, pure white feathers from visiting gulls and a large brown striped feather that I guess is from a buzzard's wing and poke into my buttonhole – there have been two buzzards circling the sky above me all morning.

Birds were exciting when I was little; they still are. The blue flash of a kingfisher on the river behind the farmhouse; a surprise visit to the heronry on my great-aunt's farm, where I watched the huge birds land and take off in slow motion from tangles of twigs high up in the trees; the day a sparrowhawk caught a wren on the back doorstep and her eyes met mine. I watched them jealously from my desk by the window at the school I loathed; they were free, I was not. I collected their feathers from the fields for my 'collection of found objects', and in the spring I roamed the hedgerows looking for their nests. Birds were part of my life, magical sky visitors, mysterious and never to be touched.

According to my phone, I have walked 14,861 steps before I spot my first and only piece of decorated Samian. It is a good-sized orange shard, about 0.6 in square, that is decorated with a curled floral swag known as a 'festoon', with a tiny rosehip-shaped bud hanging from the middle, called a 'bottle bud'. The shard is easy to spot in a pile of dark brown soil that has been dug out by a rabbit burrowing under the hedge.

The bowl it had once been part of had probably been made in the southern, central or eastern part of Gaul between the first and third century CE, where Samian, also known as *terra sigillata*, meaning 'sealed earth', was mass-produced in large kilns. The slip is slightly crazed, but otherwise it is perfect and unscratched, perhaps because it was at the edge of the field and hadn't been tumbled by years of ploughing.

The last length I walk is along another hedge. I pick some apples from a feral apple tree, possibly a survivor or descendant from when the fields were orchards. Most of its fruit has fallen to the ground and the fermenting apples smell like sweet cider. Apples, soil and leaves, these are the smells of autumn and very different to the smell of algae and river silt. There is a plum tree in the hedge too, and I pick the ones I can reach carefully because wasps are hiding in secret holes they have excavated, their leather bodies curled up and ready to sting.

I sit down with my gatherings at the end of the field, beside a clump of last year's teasels, and get out my Thermos flask. The teasels are the colour of my tea. A wood pigeon is cooing and I can hear the distant drone of a tractor in another field. Above me, swallows are lined up on an electricity wire like pegs on a washing line. A swallow can live for up to eight years and some birds can fly more than 96,000 miles in their lifetime. The swifts left in July and now it is the swallows' turn to leave, taking summer with them.

Thursday 15 September (low tide 0.67 @ London Bridge, 12.21)
Central London – South Bank

A river of people is following the Thames, waiting to pay their respects to the late Queen as she lies in state at Westminster Palace. It crosses Lambeth Bridge and snakes east along the river, past the London Eye, Waterloo Bridge, the Tate Modern and the Globe, Southwark Cathedral, London Bridge and Tower Bridge.

There is something palpable radiating from the crowd. I am not sure if it is pent-up excitement, grief or anticipation, but whatever it is, it drifts down to the foreshore along with the low hum of hundreds of voices and eyes that are watching me. People start calling down good-natured questions – 'What are you doing?' 'What have you found?' – and while I am by the wall it seems impolite not to answer, but as soon as I can, I move away from them until their voices are too far away to hear and the feeling of being watched subsides.

Near the Millennium Bridge, I count twenty-one swans on the foreshore. Most of them are pure white, but there are a few of this year's cygnets among them, as large as their parents but still smudged with grey. It takes two years to completely transform from a grey cygnet into a beautiful swan. They won't turn completely white until they are a year old, and their bills will gradually turn orange after that. I've never seen so many swans together on the foreshore in central London before, and they make an extraordinary sight. A brave woman in a bright-green dress is among them, offering them grain from a plastic bag. They waddle towards her on flat rubbery feet as wide as my palm and push and shove each other as they stretch out their long necks for food. Two weeks ago, they belonged to the Queen, but they are the King's swans now.

Richard the Lionheart is said to have brought mute swans home with him from Cyprus in the late twelfth century. They were a valuable commodity among the nobility, and the owners of swans were duty bound to mark their property with a series of nicks on top of the birds' beaks. Any unmarked swans on open waters were considered the property of the monarch, and the tradition lives on. Apart from the monarch, only three organisations can claim ownership of mute swans in England and Wales – Abbotsbury Swannery and the Worshipful Companies of Vintners and Dyers – and they ring their swans in a swan-upping ceremony that takes place annually on the Thames.

I watch the crush for food get out of control and the woman in the green dress flee to the river steps for safety. Two swans separate themselves from the group and waddle into the water. They spread their giant wings, lift up their huge bodies and begin to paddle across the surface, flapping in slow motion. Slowly, they lumber into the air, regaining something of their dignity and grace as they stretch out their necks and tuck in their feet.

There is a sound this otherwise silent bird makes in flight that I've heard on the river. It's a throbbing, whistling noise, a wing song, that can travel some distance and is made by air passing over their feathers. As they fly the length of the river, these royal birds will be singing their own sad song for Elizabeth II.

Tuesday 20 September (low tide 1.84 m @ London Bridge, 15.51)
Horselydown

There is a little passageway just east of Tower Bridge with a sign on the wall that reads: 'Private Land: part of the property of the Anchor Brewhouse'. The passageway runs through the old Anchor Brewery building, now luxury flats, to Horselydown Old Stairs and the river. The stairs are some of the oldest on the Thames and were even marked as 'Old' on cartographer John Rocque's map of 1746. I stand at the top of them, looking out over the river and up at Tower Bridge. They are deserted today, but there's often a canoodling couple hidden away here, or a photographer or painter capturing the iconic view. The tide sometimes drops low enough to walk under the bridge itself, and the view never fails to awe me. The steps are slippery at the bottom, so I take them slowly, down to the cobbled causeway that the river is breaking up and washing away. I am not completely alone on the foreshore though: a man is walking his dog and he nods a curt hello as he walks past.

Horselydown was once grazing for cattle and horses, which is where its name comes from, but by the nineteenth century the area was crowded with breweries, mills, factories and warehouses. Horselydown Old Stairs are at the end of Shad Thames, a riverside street still lined with Victorian warehouses and spanned by the walkways that connect them. The warehouses held tea, coffee and spices, and each one was dedicated to a particular import to prevent cross-contamination.

The foreshore here would have been a pungent place 150 years ago. Along with the stench of London's filth from the river and the general acrid smell of burning coal in the air, the breweries and malt houses would have mingled with the scent of spices and other odours wafting west from Bermondsey: the sweet smell of baking biscuits from the Peek Freans biscuit factory; the tang of vinegar from Sarson's vinegar works; the smell of boiling bones from the glue makers; and the pungent stench of the tanneries that used urine and dog faeces to tan their hides. The felt workers and hat makers who breathed in the effluvia of their industry – nitrate of mercury solution, which was brushed onto beaver fur to aid the felting process – went insane from the toxic effect of the mercury. The phrase 'mad as a hatter' was coined in the dark stinking streets and workshops of riverside Bermondsey.

Sometimes warehouses went up in flames. The Great Tooley Street Fire, London's largest peacetime fire since the Great Fire of 1666, broke out in a consignment of jute at Scovell's warehouse at Cotton Wharf, just down from London Bridge and close to Horselydown. It began on the afternoon of 22 June 1861 and raged for several weeks, sending a great mass of burning fat floating down the river, which set light to any small craft it reached. According to an illustrated history of the fire written shortly afterwards, 'The river seemed almost turned to blood, but so bright and lurid in its deep glow, that it actually appeared like a stream of fire.' For months after the fire had been quelled, local people were still wading into the river to skim off the floating fat, while children scooped up mud and separated the grease from it. It wasn't unusual for mudlarks and waterside residents to reap the rewards of a warehouse fire. When large boxes of tea were thrown into the river from a blazing Victorian tea warehouse, a flock of mudlarks descended with thin cotton bags to scoop up the floating tea, which they sold.

I pick up a 20p piece that someone has probably thrown off the bridge, and a small metal horseshoe enamelled in dark blue that looks as if it had once been part of a brooch or tie pin. There are the usual patches of nails, large chunks of masonry, tide-washed bricks, nineteenth-century pottery and clay pipe stems, and at the bottom of the stairs there is a gathering of small silver nitrous-oxide canisters, a fairly recent addition to the river's trove and an all-too-familiar sight these days, along with plastic vapes. Under the shingle the mud is dark grey and smelly and there is a sense of foreboding about it, as if the area's industry and poverty had seeped into the mud itself, which in a way it has. There are hidden heavy metals in Thames mud – lead, mercury and cadmium – from generations of industrial pollution and I am pleased to be wearing gloves.

I pluck a lead bag seal out of the smelly mud and wash it off in a puddle of river water. The writing on it looks Russian, which means it is likely to be a hemp-bale seal. Hemp, a variety of cannabis, was (and still is) grown for its fibre. Until the beginning of the nineteenth century, 96 per cent of British ship rigging was made of Russian hemp, but in the early nineteenth century Britain started importing jute from America, which was cheaper, so it is likely the little seal dates from before then, possibly even the late 1700s.

I carry on east as far as I can, along sand and shingle and under a walkway of river-slimed wooden posts, until I hit a dead end and can't go any further. I turn and look back at the famous bridge, which is framed by green and brown posts, and a multicoloured tangle of nylon rope that is wrapped around them.

Friday 23 September
Autumn Equinox

Today is the second equinox of the year. The sun is exactly above the Equator again and the day and night are almost equal length. According to the astronomical calendar, it is the first day of autumn, but the meteorological autumn, the fixed date used to consistently space the seasons, began on 1 September and will end on 30 November.

Autumn is a melancholy time of year. It lacks the hopefulness of spring and while it is symbolic of plenty, harvest and ripe fruits, it is also a portent of death, loss, decay and darkness. Days shorten and mornings darken, the world turns and time marches on, while tired trees prepare to shed their old clothes and sleep. Winter is coming, which means I will need to go back to my tide tables to search for low tides that fall within daylight hours.

Wednesday 28 September (low tide 1.02 @ London Bridge, 10.46)
Central London – North and South Bank

The weather has turned, there is a chill in the air this morning and it is still dark at 6 a.m. when I leave the house, but it is warm and sunny by the time I reach the foreshore, the kind of low, mellow autumn light that skitters off dust motes and magnifies colours. I feel a tickle on my forearm and see a tiny black spider clambering through a forest of golden sunlit hairs. Another is climbing the pocket of my jacket and I look up to see the sky twinkling with fine threads. Money spiders are falling through the air on gossamer parachutes that have carried them high into the sky on rising air currents, before they reached cooler air and came back down to earth. They can travel for hundreds of miles this way, clinging to their silken threads as they sail over land and sea and even up mountains. I watch her for a while, then gently blow her off my arm. Money spiders are lucky – they bring wealth and new clothes and, as it turns out, good mudlarking finds too.

My first find is a good one. I can tell it is a seventeenth-century trade token from its size and thickness, but all I can see are the words 'At the Windmill'. It needs cleaning and looking at through a magnifying glass, so I drop it into my old plastic camera-film pot to look at later. There's a glow of satisfaction I get from saving things for later and knowing there's something to research when I get home. I find another eighteenth-century clay pipe with the initials 'W. T.' for the pipe-maker William Tappin on the heel. I've found them before and know that the Tappins were a family of pipe-makers with a workshop at Puddle Dock, close to Blackfriars Bridge on the north side of the river.

I photograph the stripped bones of a bird large enough to be a duck that are as white and clean as pipe stems, then I pick up a small plastic toy baby's bottle and an eighteenth-century cufflink set with a cut-glass 'jewel' of the palest blue that would have looked very fine on the cuff of a dandy.

Next, in quick succession, I find the sawn ends of three cattle metacarpals. I've been collecting them for years and, like all the others, these have a neat hole drilled through them. I collect them because they're unusual and tactile, and once they are strung together on galvanised wire, they look like long bony spines.

Similar bones that have been excavated by archaeologists in Southwark date from the seventeenth or eighteenth century, and the holes are thought to have been part of the tanning process. Skins were delivered to the tanners with legs still attached and the bones were left on as an anchor for stretching the hide. They wasted little, and when they were finished with, the tanner sold the bones to bone workers, who cut off the ends and used the rest. Most of the bones I have, I've found on the south side of the river where there were tanneries in Bermondsey and Southwark and a Skin Market on Bankside, which is marked on Rocque's map of 1746 between coal yards, with an alley that leads directly to a set of river stairs called Goat Stairs. By the Ordnance Survey map of 1869, the market had gone and all that was left was an alleyway called Skin Market Place that was demolished with the building of the Globe in the 1980s.

In 1865, a woman called Esther Lack was living at 10 Skin Market Place with her husband John, who was a night watchman at Page's Coal Wharf on Bankside. It was an area of slum housing, behind the wharfs and warehouses that lined the river and squeezed in among ironworks, breweries, boilerworks, glass houses and other heavy industry. Esther would simply have been another poor, forgotten soul had she not cut the throats of three of her youngest children on 22 August 1865. A newspaper report into the case painted a vivid picture of the conditions she endured:

> People are unprepared to witness such a scene of misery as Skin Market Place and the human burrows near it ... Surrounding factories and manure depots made the air almost insupportable ... Heaps of filth in the street over which squalid and almost naked children played in front of their dens of homes.

Esther had twelve children over twenty years in these conditions and had lost six of them. Her husband was sick, one of her daughters was blind, Esther was almost blind herself and had been told by the parish doctor that she didn't have long to live. Upon arrest she said she thought her children 'would be better in heaven than starving about the streets'. They were the actions of a desperate, sick woman. Esther was found guilty on grounds of insanity and died two months after being admitted to an asylum.

Back at home, I get out my magnifying glass and inspect the trade token. It is similar to the White Horse penny I had found at Blackwall, but this one is smaller because it is only worth the equivalent of a farthing, a quarter of a penny. Tokens like this were tolerated by the authorities between 1648 and 1672 as a temporary solution to the lack of pennies, halfpennies and farthings in circulation. It had been issued by a tradesperson working from a premises marked by the sign of the windmill, a symbol used by innkeepers, brewers and sometimes bakers and mealmen (dealers in meal or grain). The initials of the issuer were in the centre of the token – I. (or J.) P. Tokens with a trio of initials included that of the issuer's wife, so I know that I. P. was unmarried or widowed. Women also issued their own tokens, so it may have been a woman running her own business.

The windmill pictured on the token was known as a post mill, since the body of the mill could be turned on a central post to face the sails into the prevailing wind. The business was located in 'Temple Bar Without', just outside the City of London. I look on Google Maps. There is still a road near Temple Bar called Milford Lane, which runs south towards the river from the Strand. It is intercepted by the A3211, but I see from a map dated 1670 that it once terminated at the river. The name of the road suggests a mill may have been there, but I can't find any written record of one, and even the earliest detailed map of London, the Agas Map of 1561, doesn't show a windmill. Perhaps the road was named in memory of a more ancient mill that had long since gone.

Thursday 29 September
The Sword

I email Stuart about the sword I found last year. He confirms it's sixteenth century, but he has yet to find anything similar with more context. It's still in the fridge in the conservation department at the Museum of London. The museum is closing in December for a big move to new premises that will take three years to complete, and everyone is busy. The X-ray machine has broken too, so there is a backlog of objects waiting for attention. Once they have assessed the sword, they will decide if they want to keep it, in which case it will be conserved. I have a contingency plan if it comes back to me. The Department of Archaeology and Conservation at Cardiff University have agreed to conserve it, so either way it will be safe.

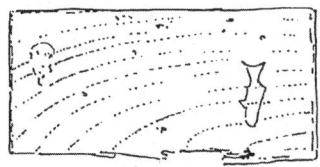 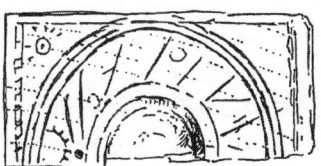

CODE: 1410.22.Q07

OBJECT:	c.17th century pocket sundial
MATERIAL:	Boxwood
DATE FOUND:	14/10/2022
LOCATION:	Central London – north bank
NOTES:	A small (3.5 x 1.5 cm) rectangular piece of wood with a circular pit in the centre, surrounded by incised circles that are marked at regular intervals. The lid is missing, but there are remains of a small iron pin to which a string gnomon would have been tethered. A small fish, stamped into the reverse, might be a maker's mark, but experts at the Worshipful Company of Clockmakers are unsure. Surface find.

OCTOBER

Central London – North and South Bank

I rescue a soggy bumble bee clinging to a tennis-ball bobbing about in the shallows. I pick her up and leave her to dry out in the sunshine on the river stairs, hoping a crow doesn't find her before she has a chance to fly away.

Today is the latest tide I will be doing for a while; the clocks are going back soon, after which my low tides will need to be no later than 2 p.m. for me to lark the whole tide in natural light. My finds today are varied, although there is rarely a tide that doesn't bring something unexpected. A live bullet, but this one is German, a 9 mm Luger from the First World War, which is rare for the Thames. The number '15' stamped onto the base of the casing means it was manufactured in 1915, and 'S' stands for Spandau Arsenal, where it was made. The phrase 'Spandau ballet' is said to have come from the 'dance' that Allied soldiers did to avoid the bullets as they crossed no man's land.

It's not unusual to find bullets on the foreshore. Someone once told me when their grandfather came back from the war, they were told to get rid of all loose ammunition and war souvenirs, which they did, over the side of the boat as they were travelling up the Thames, so that might account for some of it. When I find bullets, I usually throw them back into the deepest part of the river rather than leave them on the foreshore and I send this one tumbling into the water.

The tiny medallion I find next, in a low puddle of water, might even have been worn by one of the men advancing through the bullets. I can see the head of George V, but it takes me a while to work out what is written on the other side. It is only under a magnifying glass at home that I see it is the Lord's Prayer. Without the king on it, it would have just been a prayer token or charm, but the king's head suggests something more patriotic. It could have been a coronation souvenir, but it would also have been an inconspicuous medallion for a soldier to keep close to him on the battlefields of France.

It's tempting to read history into finds, to imagine a plain glass bead once graced the neck of an Anglo-Saxon woman, that a piece of cut bone is a prehistoric artefact, to read shapes into random pieces of lead or imagine a loop of wire is something other than just a loop of wire. It relieves the frustration of not knowing what something actually is, it brings objects alive, makes them more special and important than they are, and in some ways justifies the hours that were spent looking for them. But attaching a made-up history to an object when there is no documented proof is dangerous. It distorts the object's true past and sends misinformation out into a world that craves fast facts, whether they are true or not. I find similar medallions online, but apart from the assumptions of sellers on popular auction sites, I can find no actual evidence it had been produced for a soldier or that it had been worn on the Western Front. It is just a

Lord's Prayer talisman that was made some time between 1910 and 1936.

I have, however, found written evidence that the Charles II sixpence I also find today was bent for a reason. People have been bending coins as part of a vow for centuries. In the twelfth century, a monk from Durham bent a coin and dedicated it to St Cuthbert after he injured his testicles in a riding accident, promising to make a pilgrimage to Cuthbert's shrine if the saint would help him. People also bent coins over 'dead' children, who miraculously came back to life, and sailors bent coins in the midst of storms and prayed to their patron saint, St Wulfstan, to be saved. St Wulfstan also calmed a woman 'in the grip of insanity' when a bent coin was tied around her neck. Bent coins are said to have cured wounds and blindness, and in the thirteenth century a coin bent over William Cragh, who was hanged for thirteen counts of murder and arson, brought him back to life and he lived for a further fifteen years.

After the Reformation, coin-bending lost its religious associations and became a way of showing love and devotion. In 1557, a protestant martyr called Alice Benden gave her brother a 'bowed shilling' as a reminder of her, and in 1598 Thomas Kennet gave Bennet Dunnye a sixpence, 'bending yt once, as a token and pledge of the bargen and promis passed between them'. The giving of a bent coin increasingly became a way to promise faithfulness and betrothal, and even wealthy people bent coins for each other. In 1715, Lady Bridget Osborne, eldest daughter of the 2nd Duke of Leeds, scandalously gave the Reverend William Williams a gold coin which she 'had almost bent double with her teeth'. My bent coin is the tenth I've found in the Thames. I had the first one I found made into a charm for Sarah, and it continues its journey through time as her lucky talisman and my symbol of love.

I pick up half an eighteenth-century delft ointment pot, a late sixteenth-century cloth seal, a couple of nineteenth-century trouser buttons and what could be part of a medieval pilgrim badge. It seems like a lot, and I start to wonder: will it ever run out? I suppose it depends on what that means. Every day we are adding something new to the river's trove, even though most of it is plastic, but the number of antiquities is diminishing and one day they will be exhausted. Mudlarks, dealers and collectors have been predicting this for at least a hundred years and bemoaning the end of a 'Golden Age'. The question is, when exactly was the Golden Age of mudlarking?

Those who were mudlarking before the social-media boom might say it was better in the first decade of the 2000s, when the Clippers were eroding out finds but it was quieter and there was less competition for what was being revealed. The detectorists and diggers of the 1970s, 1980s and 1990s might argue that theirs was the Golden Age. They were the first to 'break soil' and were mostly unhindered by restrictions. Even Ivor Noël Hume predicted the end of a Golden Age in the 1950s, saying: '... it was possible to walk along the shore and expect to find at least fifteen or twenty objects that were worth retaining ... however, it is possible to go over that ground half a dozen times and find nothing'.

As far as G. F. Lawrence, otherwise known as Stoney Jack, was concerned, the Golden Age of river finds was over by 1926, and I think he might have been right. Lawrence, a 'genial frog' of a man, was a dealer in antiquities who bought objects from river workers and workmen on building sites and sold them to collectors and even to the British Museum. He had benefited from the great Victorian programme of building improvements, which included the construction of bridges, embankments, locks and weirs, that produced some of the finest and the most prolific river finds in history, and for a time he also worked as Inspector of Excavations for the Museum of London. This conflict of interest didn't go unnoticed though and the position was abolished in 1926, sending him into early retirement, but he continued to trade in antiquities.

Lawrence was never without a pocket full of change to pay his contacts, doing shady deals down alleyways and in pubs. The workmen also brought their finds to his shop, which was described by the journalist H. V. Morton as:

> ... *perhaps the strangest shop in London. The shop sign over the door is a weather-worn Ka-figure from an Egyptian tomb, now split and worn by the winds of nearly forty winters. The windows are full of an astonishing jumble of objects. Every historic period rubs shoulders in them. Ancient Egyptian bowls lie next to Japanese sword guards and Elizabethan pots contain Saxon brooches, flint arrowheads or Roman coins...*

But the past was more than just a money maker for Lawrence, and his lament for the passing of a Golden Era was genuine. According to Morton, who knew him personally:

He had an almost clairvoyant attitude to it. He would hold
a Roman sandal … and, half closing his eyes, with his head
on one side, his cheroot obstructing his diction, would
speak about the cobbler who had made it ages ago, the
shop in which it had been sold, the kind of Roman who
had probably bought it and the streets of the long-vanished
London it had known. The whole picture took life and
colour as he spoke. I have never met anyone with a more
affectionate attitude to the past.

I'm happy with what I find today, but I doubt Stoney Jack
would have been very impressed with my handful of worn
and broken artefacts. Even the mudlarks of the seventies
and eighties would have come away with far more, but it's
a good day for me and I'm satisfied. By the time I get back
to the river stairs, the bee has gone, I have a pocketful of
stories, and another tide is rising.

Friday 14 October (low tide 0.91m @ London Bridge, 11.30)
Central London – North and South Bank

I wake up early to a pale world of whispers and webs.
Spiders have spun threads through the night that have
captured the morning mist and hung it like tiny necklaces
between teasels and dry leaves in the garden. There is still
a mist hovering above the river when I reach London. The
water is perfectly still beneath it, and everything I can see is
shades of light grey. It is as if the river has climbed out of its
basin and painted the city in its own muted shades.

From above it would tell a very different story though. Low mists like this reveal a hidden London. They settle along the lines of vanished watercourses and lost rivers and fall into dips and hollows that have been covered by houses and roads for centuries. The mist finds a vanished geography and outlines it. It settles as it would have done when the land was marshy scrub, and from the air it can be read like a giant foggy map.

When the sun finally breaks through, it burns off the mist and brings colour back to the city. The sand is covered with little footprints, three toes front and one toe back. I count forty-three crows, more than I've ever seen before, and in such numbers they are quite intimidating. There are nearly a hundred beady eyes watching me. They pick among the detritus and fight with each other over the best bits. Two larger crows attack a smaller one and it falls onto its back with its feet in the air, screeching. One of its attackers hops on top of it and begins pecking while the other pulls at its tail feathers. I wave my arms to scare them off, but it spooks them all and suddenly there are forty-three crows in the air, flying so close to me I can hear the dry rustle of their feathers as they pass. They ruffle my hair, squawk and scream, flap and tumble over each other and when they are gone, an empty silence falls over the foreshore.

I pick up a cut-glass tumbler that must have been thrown in the previous night from a boat or by a drunken reveller who had left a party with their drink. It is a nice glass, just a bit dirty, nothing the dishwasher can't clean up when I get home. Where it had been lying in the mud, I notice a scattering of very new-looking screws. I know of two people who scatter screws, brass washers and copper nails over the foreshore. They do it to confuse metal detectorists, and I wonder if they have been at it again.

People sometimes plant things on the foreshore to confuse mudlarks. Not so long ago, there were mysterious white clay figurines turning up, and a few years back too many well-knapped flints were being found close to one of the bridges for them to have been genuinely prehistoric. Recently, I was speaking to another mudlark who was wondering if someone was scattering Roman coins, because so many had been found in a fairly short period of time. A lot of fake Henry VIII coins were found along the same stretch around the time of Covid. Perhaps the same person had scattered those too. While these could be seen as harmless pranks, less so the mudlarks who intentionally plant objects they have bought or acquired. It might sound far-fetched, but it has happened, and people have been caught out.

Today, there is a stolen wallet in the river and the testicles from a broken dildo. I check the wallet. It's not unusual to find stolen bags, purses and wallets in the river. They are usually missing the cash and credit cards, but sometimes it's possible to get what's left back to the original owner. This one is empty; I don't touch the testicles.

A stolen wallet and a pair of pink plastic balls isn't much to show for three hours, so I am pleased when I finally find a little rectangle of what looks like bone or ivory, about the size of a domino. It has a circular pit in the centre, and when I tip it to the light I can just see lines incised in circles around it. They are marked off at regular intervals and there is a small sun in the corner. I recognise it instantly as part of a pocket sundial because, although they are quite rare, it is the second one I have found. My other sundial is made of ivory and was produced in Nuremberg in the late sixteenth or early seventeenth century. On first inspection I assume the little brown rectangle in my hand is also ivory, or possibly bone, but when I test it for hardness with my fingernail, it is too soft to be either.

It turns out to be wood, possibly boxwood, which is tightly grained and was used to carve small and delicate objects like combs and beads. Wood can survive for centuries in oxygen-free mud, so it could be as old as my other sundial and would need to be kept in the fridge in a pot of cooled boiled water until it could be conserved properly.

The circular pit would have held a glass-covered compass, used for alignment, and a string gnomon would have been tethered between a small iron pin just below the compass and the lid. The lid is missing, but I can see the rusted remains of the tiny iron pin. The decorative lines have been scored into the wood and there is a surprise on the back. A miniature fish, so small I almost miss it. It looks as if it had been stamped rather than scored and is probably the mark of the maker.

The invention of the sundial was really the invention of modern time. It divided night and day into equal segments that have progressively obsessed and enslaved us. The antiquary Robert Hegge described it well when he wrote in 1630:

> A Dial is the Visible map of Time, till whose Invention 'twas follie in the Sun to play with a shadow. It is the anatomie of the Day; and a scale of miles for the journie of the sun. It is the silent voice of Time, and without it the Day were dumbe…

Pocket sundials wouldn't have been very good at telling the time though, and completely useless after sundown. They were more likely to be symbolic trinkets, reminders that life and time are fleeting, and that God is as unpredictable and arbitrary as the time told by a sundial.

Monday 17 October (low tide 0.72 m @ London Bridge, 12.10)
Deptford

We had a stupid argument over nothing this morning, words got taken out of context and what began as a stupid bicker blew up into something much larger than it ever deserved to be. I should have apologised. I knew in my heart it was petty and time-wasting, but I'm stubborn and instead I have gone mudlarking on an argument and that's never good. I travel to London under an angry cloud and begin at the mouth of Deptford Creek, where I know it will be quiet. I hope my mood will improve.

I hop over railings and walk straight down to the river, crunching through deep patches of rust, old nails and a surprising number of sardine-tin keys. Near the water, the sand is so soaked through with old oil it is claggy and sticky. It sticks to my boots and the warm reek of engines and heavy machinery drifts up with every step. There is still a light mist over the river, but in the distance the sun has come out and the tall buildings in the City gleam white where they catch the light. At the mouth of the creek the mud is thick, smooth and very smelly. It rained last night, and water is making its way into the river in any way it can. It is seeping out of cracks in the wall and gushing from drains. It looks as if the river wall is weeping, and this matches my mood.

Deptford Creek is not the name of a river; it is the name of the tidal mouth of a tributary of the Thames called the Ravensbourne. It is always mucky and rubbishy around Deptford Creek, but sometimes old bottles or long, well-preserved pipes emerge from the mud. Silt, washed down the creek by the rain, has mostly covered the foreshore today and there isn't much more to find, so I climb back over the railings and take the river path over the creek to Watergate Street. In the eighteenth century there were three sets of stairs here – Upper Watergate, Middle Watergate and Lower Watergate. Only one set remains, Upper Watergate Stairs, at the end of an ancient alley that's paved with unusually large cobblestones that are thought to have come by ship in ballast and may even have come from the other side of the world.

The concrete stairs deliver me straight onto an old bargebed, which is in good condition, with its cap of chalk still intact. A man walks past with a yellow dog, the only other person I've seen on the foreshore, and I ignore him. There's still a black cloud hovering over my head and I'm not in the mood for polite conversation. As I step off the chalky patch I spot a white ball, about the size of a small plum, and pick it up. It is made of white pipe clay and has clearly been made by hand, as it is slightly irregular and I can see fingerprints in the fired clay. These white balls, or 'rattlers' as they are known, because they are hollow and often rattle when they are shaken, mostly appear on this stretch of the river and are something of a mystery.

I don't think they are milling balls for mixing, grinding or polishing because those are usually solid, but they might be paint-agitator balls that stopped the paint in tins from settling. An old mudlark I know has found 1,200 agitator balls at Wapping and has traced their origins to an old East End paint factory, where they were used to stir paint. The ones he finds are a variety of colours – pink, blue and occasionally green – having taken on the pigments of the paint they once mixed. I have found a few myself, but they don't look like the Deptford rattler balls. Someone once told me they were used to test the temperature of kilns before the pots went in, but I'm dubious of that theory too. I know balls of pipe clay were used to touch up marks on the white breeches of soldiers in the early 1800s, but these balls are fired and wouldn't have been much use for that either. The most plausible theory I think is that they are part of a modern art project. It's not unusual for artists to 'release' their work to the river. On Easter weekend in 2012, the artist Anon released 5,000 numbered ceramic eggs into the river that have so far been found as far east as the Essex coast.

I walk under a huge derelict concrete jetty and past a floating iron platform that groans and creaks like an old ship as it rides the waves made by passing Clippers. Near the river wall I see an old pram wheel, some car tyres, an ancient Nokia phone, the iron dome from an alarm bell, lots of bricks and a few old shopping trolleys. The tide has turned, and as I walk, I keep an eye on the long ladders that are stationed at intervals along the river wall. There are pinch points here and I want to make sure I can escape if I have to. The only other set of stairs is Drake's Stairs, which are always locked.

Drake's Stairs once led to one of Henry VIII's docks, Deptford Dock. He established two naval dockyards on the Thames, one at Deptford and one at Woolwich, a few miles east of Greenwich. By the end of his reign, the yard at Deptford had grown to be the more important of the two, but there is little left of it now. In its time the Great Basin held the *Mary Rose*, Henry's famous flagship that sank into the silt of the Solent in 1545. It also laid up Sir Francis Drake's *Golden Hind*, which circumnavigated the world between 1577 and 1580. It fitted out Elizabeth I's ships for the Armada in 1588, as well as those for James Cook and George Vancouver's voyages of discovery in the eighteenth century. Deptford Dockyard prepared vessels for Nelson's battles and it served as a military base well into the twentieth century. The mud here is soaked in naval history.

Before I reach the thin sliver of sand that runs along the wall to Drake's Stairs, I stop at an area filled with old leather shoes and shoe soles. Most are heavily studded with hobnails and very worn. They look nineteenth century, possibly even early nineteenth century, and I wonder if they are anything to do with the dockyard, perhaps worn-out military shoes and boots that had been dumped in a bargebed when they were filling it up. I keep two inner soles, one right and one left, and a studded, right-footed outer sole.

Near the shoes is a patch of sand filled with small bits of metal. A perfect hunting ground. I find the usual dress pins and among them a delicate copper-alloy cravat pin, with a simple stylised flower for a head. According to a jewellery expert I show it to later, it is likely to date from around

1860, by which time the wearing of cravats had become popular among all but the working classes, and affordable cravat pins in lower-quality materials were being produced to meet the demand.

I get three steps along the narrow strip of sand to Drake's Stairs, and just in time, I hear a boat coming. I sprint the remaining distance and narrowly avoid a swamping. The wide algae-covered steps seem out of place, faded grandeur in a neglected and forgotten setting. They rise up from a crumbling causeway and lead to a high ornate iron gate, decorated with a pair of gilded crossed anchors. Someone has managed to dump a yellow hire bike over the gate, and it lies in a tangle at the top of the stairs. I look back to the giant ugly jetty. The tide is coming in fast and the wash of passing boats is already covering my escape route. I'm reluctant to leave and I'm still feeling stubborn, but I can't delay it any longer; it is time to go home and apologise.

Wednesday 19 October (low tide 1.65 m @ Ramsgate, 13.24)
Kent – Broadstairs

I was planning to visit the river today, but plans change. I need to be at home to collect the children from school, so I slip down to my favourite local beach for an hour instead. I avoid beaches when they are crowded with visitors, and have barely been all summer, but it is safe to return now that the holidaymakers have gone, and the beaches are empty again.

I want to look for sea beads, which I've also found on the Thames foreshore. These small, round, creamy-white beads are not manmade, although they are often mistaken for ceramic beads. Sea beads are small Cretaceous fossil sponges, *Porosphaera globularis*, that have been naturally drilled by the sea through the softest part, the central cavity and the osculum, which in life was used to expel water. Most of the little sponge fossils I find are still perfect or only partially drilled, but about one in ten has a hole all the way through and these are the ones I collect.

I am by no means the first to collect sea beads. In the nineteenth century, a Bronze Age necklace was excavated from a site at Higham Marshes, between the Thames and the River Medway. It was found around the neck of a crouched skeleton in a stone-lined grave and consisted of seventy-nine *Porosphaera* beads. Around 1920, archaeologist Herbert Toms wrote about them in relation to his specific interest in local folklore, noting 'it was common practice for women in Brighton to wear a solitary *Porosphaera* on a string or tape around the neck in order to ensure good health'. Toms recorded one enterprising lady selling strings of these lucky beads opposite the old pier, and beads being worn for luck at Newhaven in Sussex. He also spoke to a man called Mr Henson who, around 1850, had seen a string of beads hanging over the mantelpiece of a farmhouse in Theale near Reading. He said they were considered lucky and had been handed down in the family as an important heirloom.

I found another twenty-four beads today, but I'm not sure if anyone will ever want my strings of *Porosphaera*, let alone treat them as an heirloom. Perhaps I'll be buried wearing a sea bead necklace and take them with me.

Friday 21 October (low tide 1.35 m @ London Bridge, 17.50)
Central London – South Bank

It is a terrible tide today. The water stays much higher than the tide tables predict, probably because it has been raining a lot further upstream, then I lose the light when dark storm clouds descend and block out the sunset. The foreshore is papered with orange, yellow and brown plane-tree leaves that flop and float in the shallows and cover all of my best spots. As I scrape some of them aside with my foot, I think about the trees along the Embankment at Westminster and the green leaves that shaded me in the summer that are now floating past. In time, many of them will wash up on the Isle of Dogs' great bend, where they will gradually turn to mush and reveal the banknotes, plucked from fingers and blown into the river, that are hidden among them. While modern coins will rust away, in centuries to come people will still be finding our plastic currency.

A swan is preening itself among the floating leaves in the shallows. Its neck looks oddly flexible and long, like a pipe cleaner with a tiny head on the end. It chatters its orange beak through its feathers, shakes, ruffles, and dips under the water, waggles its tail and flaps its long wings. Then it twists its neck and rubs its head into the feathers on its back and against the waxy gland near its tail. Once we had singed away the last bits of down with matches, we used to light the tail glands of the ducks we plucked on the farm when I was a child. They would burn for a while, like pink fleshy oil lamps.

I watch the swan as it scatters downy feathers across the water and on top of the leaves. The wind blows some of them onto the mud, then a single large wing feather, pure white and perfect, slowly floats ashore. I wait for it at the water's edge and carefully tuck it inside my jacket, hoping it won't get crushed as I mudlark.

Medieval quills were typically made from the first five flight feathers from the wing of a goose or swan. Left-handed quills were made from feathers from the right wing and right-handed quills were made from the left. I think mine is a left-wing feather. A good quill needed to be hard, dry and clear with age, and from the eighteenth century this was done by heat curing in a process called 'dutchifying' or 'dutching', because it was first done in Holland. By the 1830s, a simplified form of dutching had been developed that involved soaking the feathers in water for several hours and plunging them into hot sand or ashes. The hardened quills were sold in small bundles to be cut later, or they could be purchased from pen-cutters ready to use.

Apart from the feather, finds today are predictably scarce. I hold my breath as I pull a chunk of mosaic out of the mud, but it isn't a piece of Roman floor. Twelve neat cream, red and maroon blocks are embedded in concrete and not Roman mortar, which generally has large inclusions of terracotta mixed in with it. It is probably Victorian or Edwardian, when it was fashionable to decorate the entranceways to shops, pubs and grand houses with mosaics.

The late nineteenth-century tailor's button isn't from the usual cheap East End tailors I generally find. It has the name 'Bingham, Conduit St' on it, Conduit Street being just off Savile Row in Mayfair. It may have popped off the fly of a wealthy gentleman and washed down more salubrious drains into the river, or it could still have come from a river worker. Clothes of the wealthy often ended up on the backs of the poor through donations, gifts to servants and eventually the rag markets of the East End. But my favourite finds of the day are four broken and blackened hazelnut shells that I discover under the leaves in some freshly eroding mud. I've found them before; they don't last long once they've been picked out by the river, but if you catch them in time, they dry well.

Hazelnut shells were discovered by archaeologists excavating the site of Shakespeare's Globe, close to the Thames on Bankside. They initially assumed they had been dropped by the audience who were thought to consume hazelnuts like popcorn, but more recent theories suggest they were a component of a type of non-slip, free-draining flooring that included ash, cinders, and sand and silt from the Thames.

When they built the new Globe, they decided to lay the same mixture in the yard and imported 7.5 tons of broken hazelnut shells from Turkey, but according to a front-of-house show report dated 29 May 1999, it didn't go entirely to plan. The weather that day was 'everything except snow' and the report described the drains being blocked by feathers, straw from the thatch and hazelnut shells. The sixteenth- and seventeenth-century theatres on Bankside would have had a very different drainage system and probably wouldn't have had the same problems, but I'm sure hazelnut shells still found their way into the river.

Sunday 30 October
Clocks Go Back

The clocks stepped back today at 2 a.m. and we will pay for our extra hour of sleep with darker evenings. Winter is here.

As the days shorten and nights lengthen, time overlaps and the thin membrane between this world and the next becomes stretched and thin. Tomorrow is Samhain, All-Hallows Eve or Halloween, when the lives that have been poured into the river over millennia rise up from the mud for a night. Fishermen, lightermen, sea captains and scavengers will push through the skin that divides them from us and for a brief time we will walk side by side, looking out over the tide together.

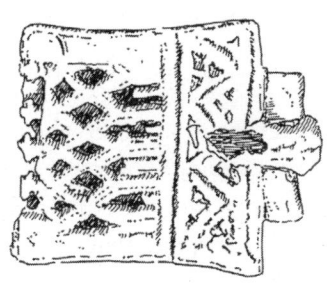

CODE: 3011.22.SB10

OBJECT:	c.14th century buckle plate
MATERIAL:	Pewter
DATE FOUND:	30/11/2022
LOCATION:	Central London – south bank
NOTES:	A medieval buckle plate, decorated in gothic style and made for a one-inch leather belt. The buckle frame and pin are missing. A similar buckle can be seen in the painting *Portrait of a Lady* by Rogier van der Weyden (c.1460). Surface find.

November

Wednesday 2 November (low tide 1.77 m @ London Bridge, 14.05)
Central London – North and South Bank

There is already a man at my spot when I arrive today, and he is methodically scraping away several inches from the top layer of the foreshore. I wonder if he's noticed that he is also scraping through the sewage that has been left as the tide fell. Among the leaves that have blown off riverside trees in the previous night's storm, there is a multicoloured scum of microplastic spread like jam over the foreshore, along with the usual nasty things that wash up after a sewage spill: ear plugs, tampon applicators, cotton buds, polystyrene, pill packets and bottle caps in all colours. There is also quite a lot of semi-digested sweetcorn, tomato skins and the perfect crescent moon of a nail-paring – all grim reminders of why I wear gloves and why the man frantically scraping at the foreshore should really be wearing gloves too.

Since I started mudlarking, I've watched the amount of sewage in the Thames increase every year. Until this year I'd never seen such fresh raw sewage washing up on the foreshore, but on a recent visit to Greenwich I was literally tiptoeing around turds and the stench was overwhelming. Forty million tonnes of sewage goes into the Thames each year; eight million tonnes of that is from Greenwich alone.

They are building a 'Super Sewer', the 16 m, 23 ft wide Thames Tideway Tunnel, that is set to open in 2025. It runs between Hammersmith and Limehouse and follows the course of the river up to 200 ft below it, intercepting the old sewage system, storing the waste, and transferring it to updated treatment plants east of the city. But until then we have to manage with a groaning underground Victorian network. It is shocking and disgraceful, and it sometimes feels as if we are going backwards.

The river has been a convenient repository for London's filth since the city was founded 2,000 years ago. In 1357 it was so disgusting that the king sent an order demanding it should be cleaned up. 'When passing along the water of the Thames,' he wrote, 'we have beheld dung and laystalls and other filth accumulated in divers [diverse] places and have also perceived the fumes and other abominable stenches arising therefrom…' A sanitation report of 1844 recorded coals, cinders, bottles and broken pots, as well as old hats, dead cats and scrubbing brushes being thrown down London's drains, and not much has changed. I always tell people, 'If it fits down the toilet, it will end up on the foreshore,' and it does.

On a microscopic level, scientists at Royal Holloway University in London have estimated that 94,000 microplastics per second flow down the river in places. This is stuff you can barely see: glitter, microbeads from cosmetics, plastic fragments from larger items and synthetic fibres from washing machine outflows. The Thames is facing another crisis of waste, and just because we can't see it, doesn't mean it is any less concerning.

The Victorian sewage system we currently rely on was built at another crisis point for the river. In 1842, *Punch* magazine published an eye-opening menu of rubbish and pollution flowing into the river that included: decomposed vegetation from the Berkshire and Surrey marshes, mixed with animal matter 'including a large number of drowned kittens'; 'emollient streams' from the Brentford soap works; brewery waste from Chiswick; drain flow from Hammersmith; and lime from Vauxhall. Lambeth 'poured forth a rich amalgam from the yards of knackers and bone-grinders'; Horseferry 'gave up dead dogs'; Westminster had a 'common sewer of uncommon dimensions'; the Fleet ditch bore 'the concentrated essences' of several London neighbourhoods and the Tower ditch added its 'delicious slime' from yet more pungent boroughs. Finally, the Surrey (south) side yielded the essences of industry, the 'refuse of tar-works and tan-yards'.

On 7 July 1855 the scientist Michael Faraday was travelling at low water by steamer upstream from London Bridge when he noted that the boat was surging through 'an opaque pale brown fluid', which he later described as 'feculence rolled up in clouds so dense that they were visible at the surface'. The sanitary commissioner's report of 1857 noted that 'large quantities of objectionable matter are frequently left on the foreshore to be washed away by the tide' and finally, in 1858, after a long hot summer, the smell rising off the Thames was so overwhelming that Parliament was abandoned.

It became known as the Great Stink and, eighteen days later, an Act of Parliament was passed, making a huge amount of money available to find a solution. The engineer

Joseph Bazalgette was commissioned to design a system of large sewers that took the waste to the eastern edge of the city. Although Bazalgette's sewers improved the quality of the water in central London, the problem didn't just go away, it simply moved downstream where the city's sewage poured into the river. Huge 'mud' banks built up around the sewage outlets and the river downstream became a stinking cesspool.

But this sudden surge of concentrated sewage created opportunities for the ever-resourceful river scavengers. In November 1876, *The Times* published an article entitled 'Butter from Thames Mud', in which it described mudlarks on the north shore, from Dagenham downstream, with 'basket on back and hooks in hand, busily engaged in collecting Thames fat'. The article claimed they were searching for balls of fat, walnut-sized to the size of a cricket ball, that washed up at low tide. The balls, which were said to be slate grey on the outside and white in the middle, had a cork core that was enveloped in a mass of hair and wood fibre upon which the fat collected. They were allegedly purified by successive boiling and other processes before being made into butter.

These 'fat balls' sound very familiar. They are what we would these days call fatbergs, monstrous spreading entities that lurk in the sewers under London's streets and feed off congealed fat from household kitchens and restaurants. Sometimes they break up and I find pieces of them on the foreshore. In response to the article, on 16 December 1876, the *Sanitary Record* questioned the use of the fat as butter, but confirmed that they had seen it being collected by scavengers at the outlet of the sewer at Barking Creek. Four men were seen skimming the surface with nets and collecting their finds in a boat until they had enough to take to a nearby barge that had equipment to extract and purify the fat. The *Sanitary Record* assured the public that the resulting fat was only good enough for soap and candles and wouldn't be turning up on their breakfast tables.

Fat was a popular quarry for mudlarks and they didn't only search for it at Barking, they also skimmed the river further upstream in London. In the mid-1800s Henry Mayhew interviewed an 'Irish lad of about thirteen years of age' who he found gathering chips of wood in an old basket at high tide. The mudlark told him he also collected logs, pieces of iron, rope, copper if he was lucky, and fat, explaining: 'We also pick up pieces of fat along the river-side. Sometimes we get four or five pounds and sell it at 3/4d. a pound at the marine stores; these are thrown overboard by the cooks in the ships, and after floating on the river are driven on shore.'

In 1887 the Metropolitan Board of Works began to filter out the solid waste and to load the sludge into six specially designed vessels that dumped it in a particularly deep part of the outer Thames Estuary known as the Black Deep. It took twenty minutes for the waste to flow out once the valves were opened and the thick, dark stain it made in the water earned the vessels the nickname 'Bovril Boats'. This practice continued until 1998.

I only touch what is necessary on the foreshore today, and pickings are slim. I give a couple of pipes to a lady who has come all the way from Finland to mudlark, advising her to wash them carefully. The north shore seems less affected by the sewage, and I manage to find a small handful of

Roman pottery shards, including some that are so thin and delicate they look as if they had been made yesterday.

I've found some weird and wonderful things on the foreshore over the years: lots of false teeth, hearing aids, crutches, a box of human ashes, a glass eye and even a human fingernail with flesh still attached to it. My most revolting and shocking finds today are the corpses of eleven fat, naked rats. Their fur has been stripped off by the river and their skin is rinsed clean, papery thin and wet. They look like white, ethereal beings from London's underworld, with bloated bodies, wide black eyes, stiff legs and bared teeth. I am so taken aback by the first rat that I almost step back onto another, and when I look into the distance, I see small white bodies washed up all along the foreshore. I've never seen anything like it. They must have drowned during a storm when the sewers flooded. Perhaps the rat population beneath London has been growing through the drought and the storms are clearing out the excess population.

As the sun prepares to set, the dead rats glow white in the silvery, crepuscular light, and I watch as the tide rises and takes their little bodies one by one.

Thursday 3 November (low tide 1.60 m @ London Bridge, 15.38)
Central London – North Bank

It is raining again. According to the Met Office, there are three main types of rain, classified according to how it is generated. Convection rain forms when the sun heats a patch of ground, warm air rises, and the rain falls as a shower.

Orographic rain occurs when warm air is forced upwards by a hill, cools as it gets higher and falls as rain. Frontal rain is when warm air meets a cold weather front higher up in the sky and forms clouds that rain persistently all day. Today's rain is frontal rain, and it doesn't let up. Not that I mind. It is mild, the rain is 'warm', and its persistence has kept everyone else away. There aren't many people willing to spend five hours with rain pattering on their backs and water streaming off their hoods, on the off chance they might find something someone lost years ago.

The rain drips in a thin line off the edges of bridges and runs in rivulets down the river walls. It sends slender sprites jumping and dancing across the river's surface, and varnishes the foreshore in a wet film that makes it easier to spot things. Where the river was frugal yesterday, it is generous today: it never ceases to amaze me how one tide can be so sparse and the next so bountiful. In the first fifteen minutes, in a patch no larger than a hundred feet square that I had thoroughly searched yesterday, I find a medieval thimble, wonderfully named *vingerhoed* in Dutch, meaning 'finger hat', that is shaped like a tiny beehive. I also find a medieval shoe sole (right foot), a late sixteenth-century cloth seal with a heart-shaped monogram, and a 'curious thing'. It is about the size of a button, silver and engraved with leaves on one side and gold on the other, with a sprung catch and an engraved sun with the name 'West' beneath it.

It looks like something from a clock or watch, but a bit of research when I get home reveals it to be a solitaire or bachelor button. The one I have found had been patented by George West and sold around 1870–90 as an emergency repair button. Half of mine is missing, but it would have had two winged projections that, when depressed, released their hold on a central shank so that the back and front of the button came apart. It could then be attached to any thickness of fabric before being snapped shut again. Ingenious.

On the north side of the river there are a couple of other waterproofed figures hunched over the foreshore. I find a musket ball and a rectangle of bone with the initials 'T+I' scratched into it. It may have been a love token or just someone doodling on a bit of bone a few hundred years ago. Then I snatch a jeton out of the waves. I can see it's late sixteenth century as I balance it on my forefinger, but suddenly a gust of wind blows it away. I hear a light ting as it hits the gravel and I fall to my knees to look for it. I try to think like a lost thing. I imagine myself tumbling through the air, flipping and spinning in the wind as I fall, landing and bouncing once, twice, then sliding between the wet pebbles, away from the searching eyes above.

It can only have landed in a small patch near my feet, but I can't for the life of me find it. Another mudlark joins in the search. I tell her she can have it if she finds it, but even with two pairs of eyes on the ground, we can't find it. It was mine for less than a minute before it went back to the river. I suppose some things just aren't ready to be found.

Tuesday 15 November (low tide 1.43 m @ London Bridge, 11.32)
Central London – North and South Bank

Today's rain is drizzle, a soft, gentle soaker not even enough to call rain, that drifts down from a low white sky and casts a wet rug over the foreshore. First up is a small iron cannon ball and a plastic chain from a Lego set that I recognise from the twins' multicoloured pile at home. A complete set of sheep teeth, still embedded in its jaw, is chomping its way to the surface and next to it is a nineteenth-century watch-winding key.

The rain picks up about two hours into the tide and every so often a trickle of water suddenly runs off the peak of my hood onto the ground where I am looking. Then I feel it leaking in at the seams of my trousers again. I still haven't bought those new waterproofs and I am getting wet, but another silver ring makes up for my soggy jeans. This one is plain and quite chunky, but there are no hallmarks on it to suggest its age. I find it in almost exactly the same spot as I found the sixteenth-century posey ring in May, and like the posey ring, it has also been cut open. Quite a coincidence, but I could never prove they are connected.

Friday 18 November (low tide 1.62 m @ London Bridge, 14.56)
Central London – South Bank

I am on the foreshore in my good shoes and smart clothes again. I have been asked to do the 2022 Canon's Lecture at Southwark Cathedral, which is a great honour and perfectly timed to catch the low tide before it starts. I tuck my trousers into my socks and skip around the mud carefully, I don't want to inflict the gentle waft of river mud on a cathedral full of clergy.

My only find of any note is a rusty padlock. It is misshapen and knobbly from the pieces of shingle that the rust has engulfed like an amoeba, creating a crust of pebble-dashed armour. It is rather beautiful, but the question is, should I leave it as the artwork the river had intended or risk knocking off the rust to see if there is enough sound metal to save the padlock? Before I can decide what to do with it, I have to get it home.

It is too dirty to go straight into my bag and I don't have a plastic bag with me. I do have the clean handkerchief Sarah put in my pocket before I left, though, so I decide it will have to do. I wash the padlock off, reaching out between waves and keeping as far back from the water as I can, then shake it vigorously to get it as dry as possible before wrapping it carefully in the handkerchief and putting it in the capacious pocket of my mac.

I will live with the padlock for a while before I decide what to do with it and I won't get it out again until I get home, because I never do that. Sometimes I wait for several weeks or even more than a month before I look at what I've found on the foreshore. It's like waiting to unwrap a present. While they are in my bag and unwashed, they are in a limbo world between lost and found, still pure and untouched by anyone but me. I can make the thrill of the find last as long as I want and when I finally unpack them, I need to be in the perfect place – at a table, in a quiet house, with time ahead of me to concentrate.

Three undisturbed hours on a rainy afternoon with a bag of muddy finds is my idea of bliss. When they are clean, I like to live with them for a while longer before I put them away. I leave them around the house and balance them on books on my desk, line them up under my computer and keep them on my bedside table, so that I can look at them, pick them up and think about them some more. Even then, I'm still possessive. Much to Sarah's irritation, I don't like anyone else touching them until I'm ready, because once I let them go, each person who handles them will take them a little further away from the moment I brought them into the world from the mud.

At the top of the stairs, I untuck my trousers and brush out a thin line of coarse sand that has somehow found its way into my turn-ups and scuff the mud off my shoes on the pavement as I walk to the cathedral.

Monday 21 November
Salisbury

'I'll bring you back a dinosaur poo,' I shout to the twins as I shut the front door. I have been planning this trip for months: a night away staying with an old school friend, a visit to Lyme Regis for the fossils and to Salisbury Cathedral to take my fifteenth-century pilgrim badge back to St Osmund's shrine.

It is raining when I leave and it is raining when I arrive in Salisbury. It falls almost horizontally and soaks me underneath my umbrella before I even reach the Chapter Office, which is in a complex of beautiful old houses opposite the cathedral. Safely inside, Cathryn, the dean's PA, leads me down a wonky corridor to the dean's office, where he is waiting with Emily, the cathedral's archivist.

We chat briefly about the ghastly weather and Cathryn brings me a cup of tea, then we get down to business. I get Osmund out of his wrappers and hand him to the dean.

'How do you know it's Osmund?' he asks. I explain how it has been seen by the British Museum and that it matches another badge in Salisbury Museum so closely that it is thought they were made from the same mould. We talk about saints and shrines and whether or not the badges were given to the river as a sacrifice. 'Whatever Chaucer says, pilgrimages were dangerous,' says the dean. 'It wasn't about the journey, it was about getting to the shrine, saying your prayers, being as close as you could to God, then getting home safely without being injured, getting sick or being attacked by bandits. It would have been a great relief to return safely, so throwing a badge into the river would have been a way of thanking God for that.'

Osmund was a fairly minor saint, as saints go. He was known for his holiness, but he was a pretty unextraordinary character, a nephew of William the Conqueror (r. 1066–87), and more of a holy man with a gift for administration, not a martyr. 'Let's call him a B- or C-list saint, rather than an A-lister like Becket,' says the dean. He died in 1099 and was made a saint in 1457, which wasn't long before the Reformation in 1534, when the veneration of saints was denounced. This meant there was a relatively small window of around seventy-seven years in which my badge was probably made.

They trundled Osmund's stone sarcophagus down the hill from the old cathedral at Old Sarum to the new cathedral they had built in Salisbury from 1220 onwards, and when he was canonised, they built him a new stone tomb with arches, called foramina, on both sides so that visiting pilgrims could get their ailing arms and heads as close as possible to his remains, in the hope of a cure. 'Saints made real the presence of God on earth, so being next to their earthly remains was a meeting place of heaven and earth, as close as anyone could hope to be to God in their lifetime,' says the dean.

There would have been an impressive wooden structure on top of the stone tomb that was covered with jewels and gold, but it was dismantled after the Reformation. 'It took several days to prise out all the jewels on Osmund's shrine,' says Emily. 'They probably went to replenish the king's coffers.' They may have got rid of Osmund's body at the time too, but there is a chance the priests knew what was going to happen and buried his bones somewhere about the cathedral. Either way, Osmund's earthly remains have been lost to time.

Emily offers to take me over to the cathedral, but before we strike out over the Green in the pouring rain, we visit the archive, where the cathedral's old manuscripts are kept in a strongroom, behind an old, heavy, safe-like door. 'The first documented miracles are from around 1160 – a diseased jawbone, a "rupture", paralysis, insanity and blindness,' says Emily. 'One man accidentally sat on the shrine and was struck down with a headache. It wasn't until he returned to the tomb and prayed for forgiveness that it went away.'

Emily has laid out an ancient book for me to see. It has crinkled vellum pages and perfect, curled, oak-gall-ink writing, and I wonder if the scribe had used a swan or a goose feather and if he was left- or right-handed. The book was written in the fifteenth century and has the snappy title *Registrum in causa canonisatiotis beati viri Osmundi olim Saresbirien epi in Anglia*, or *Register of the Canonisation of Saint Osmund* for short. It contains the accounts of witnesses to various miracles attributed to Osmund when the application for his formal canonisation was being made to Rome. A paralysed man was laid next to the tomb and could walk again, a woman having trouble breastfeeding prayed to Osmund for help and found she could suddenly breastfeed her child. Another woman called Julia is described by witnesses as being 'mad'. She was spitting and tearing at her clothes, but after spending five nights sleeping by the tomb, she was cured. Drowned children came back to life, toothaches were cured, stab wounds were healed and blind people could see again. Osmund was indeed a miracle worker.

We leave the archives and walk across the green to the cathedral. It would have made sense for the badge sellers to have set up their stalls there, to catch the pilgrims as they made their way to the cathedral and queued for the shrine. As we round a corner, we are hit by a strong gust of wet wind that almost knocks me off my feet and turns my umbrella inside out. 'You can see why they call it Kill Canon Corner!' shouts Emily. Inside, the cathedral is cavernous and awe-inspiring, the architecture designed to draw the gaze upwards to the heavens and to inspire reverence, especially in those who lived most of their short lives in small, dark, smoky dwellings.

I walk past recumbent stone bishops, dukes and earls, their palms pressed together in prayer. They have swords by their sides and dogs and lions at their feet. Some are missing noses and toes, and there are centuries-old initials scratched and carved across their cheeks. I look carefully at their chain mail, buckles and sword chapes, all of which I've found on the foreshore.

Osmund's shrine is in a corner at the east end of the cathedral. It is very plain and unassuming. There are three fat, bean-shaped arches on both sides and the grey Purbeck marble is worn smooth and shiny from millions of hands. I take Osmund's badge out of its plastic box, unwrap the tissue paper and, looking around to make sure I'm alone, I kneel down on the cold floor beside invisible medieval pilgrims. I can hear their whispered prayers and smell the sweat and dust of their journey. Their words echo up into the great space above the tomb, twisting around each other in a mist of devotion, gathering in the rafters, squeezing between joists and settling into the patterns on the carved stonework. Shoulder to shoulder with these ghosts of the past, I insert my hand through the middle arch and gently touch my badge against the empty tomb.

Tuesday 22 November (low tide 1.19 m @ Lyme Regis, 10.40)
Lyme Regis

Driving into Lyme Regis, I am met by a sign telling me I have arrived at the 'Fossil Coast'. There are cast-iron ammonites on the lampposts and, as I wind my way through the wet, narrow, empty streets, I pass fossil shop after fossil shop. It's like driving through a theme town and I feel a little dispirited. I park the car and squint through the pouring rain for Mike, a professional fossil hunter who spends most days searching here. We follow each other on social media, but we have never met. A while back I asked if I could come with him one day and he happily agreed. Now I am looking for a total stranger in a deserted car park in an almighty deluge of rain.

I see a solitary figure in the distance and assume it's Mike. His clothes give him away. He is dressed for the weather in muddy waterproofs, green wellies and a baseball cap, and is carrying an old-fashioned grey canvas rucksack that I can tell is already heavy with what I assume are chisels and hammers, the fossil hunter's toolkit.

Despite the weather, Mike is a ray of sunshine, and my spirits lift. I can tell he is itching to get down to the beach, so we set off at a pace. 'We're a bit late on the tide,' he says as we go down the steps, 'so we'd better get going'. Generally speaking, autumn, winter and early spring is time to collect, and the bit in between is for preparing the fossils.

Fossiling at Lyme is best after a good high tide or a storm that washes over the lumps of rock and mud that are continually falling out of the cliffs. Today's tide isn't particularly good, but I was still hopeful. 'You have to be careful with the tide,' says Mike. 'Even when it's low it can suddenly surge up and catch you out.' The rain gradually eases off and the sun tries to break through, spreading an eerie yellow light across the sea and up the beach. I look up at the cliffs where mud is rolling and crumbling off the sheer sides. 'Don't go near the cliffs, either,' says Mike. 'They're always moving, especially after rain like this, and they can slide at any point. A few years ago a woman was killed a little way along from where we were standing, and it took them hours to dig her out.'

Mike reads the cliffs in the same way I read the foreshore, but for him it is like looking at a layer cake with millions of years in each slice. He knows, when certain layers collapse onto the beach, what they are likely to contain, and he is continually scanning the cliffs for fossils that are about to fall out. My first find isn't a fossil, though.

As Mike is explaining how the ichthyosaur bones work their way out of the cliff and fall onto the beach, I spot a small cross-shaped thing next to his foot. It looks like a World War II German Iron Cross, but with no engraving, it has to be something else. It is a medal, I am sure of that, but it takes me some time later to work out which one. The central medallion, which would have had the dates 1915–18, has fallen out and it is badly pitted and eaten away by time, salt water and acid soil, but the shape and the two crossed swords confirms it to be a World War I Croix de Guerre.

This medal was created to honour acts of bravery and was conferred on any member of the armed forces, French citizens and foreigners who had been mentioned in dispatches. It was awarded to entire towns and once even given to a pigeon called Cher Ami who helped to save the lives of 194 American soldiers by carrying a message across enemy lines in the heat of battle. I wonder who had received this medal and why. It must have been thrown away, either on purpose or by accident, and ended up in the town dump.

The dump on top of the cliff had been officially used from 1908 to 1974, although it had started to fizzle out of use in the 1950s. Then, one evening in late May 2008, there was a huge landslide. According to Mike, the following morning was perfectly calm and sunny, the tide was in and a small watercourse was running from the cliff, through the freshly exposed dump material. It carried with it broken glass and bottles that tinkled as they made their way down to the sea, where scores of old bottles were bobbing about on the surface. 'It was surreal,' says Mike. Initially the landslide was dry and dusty with a 'strange earthy smell' that became more chemically pungent when it rained.

I poke around in the rusted iron, pottery, glass and old batteries and find a number '6' from a military badge; a World War I Machine Gun Corps button; the rusted remains of a pocket watch (the little cogs were scattered nearby); a handful of pitted and crusty pre-decimal coins; a green bead; and a marble. I'd been told a lot of military badges and buttons turn up here, mostly from World War I, but nobody really knows why.

In among the rust and glass I find my first ammonite. It is a perfect, dull-gold spiral no larger than my little fingernail. 'Ammonite' comes from the name of the ancient Egyptian god Amun, who was depicted with a fine set of ram's horns and this is what it looks like: a tiny, tightly curled ram's horn, balanced on my cold, wet finger. 'They're beautiful,' says Mike when I show him, 'but they're not easy to keep.' I already know its time in this world will be fleeting from bitter experience. The pyritised twigs and tiny shells I had collected from the London clay at Warden Bay on the outer Thames Estuary disintegrated into a pile of dust within four years and I had to hoover them out of my finds chest. They had taken millions of years to form and just a couple to crumble away to nothing.

We chat as we search. I find some more pyrite ammonites, then, only half an hour after stepping onto the beach, among the last century's old stair-rod fixings, broken Shippam's paste pots and dissolving batteries, I find an ichthyosaur vertebra. It is a chunky piece of grey stone, seven-sided and concave. 'That was a lucky find,' says Mike, 'there's been at least four professionals over here this morning already.' I'd seen him say a brief hello to someone earlier, and in the distance I could see a couple of intense searchers in wellies, but I hadn't noticed that many other people. 'Is there much competition here?' I ask, somehow expecting the world of fossil-hunting to be less competitive and political than mudlarking. 'Yes, we tend to work alone,' says Mike. 'It's highly competitive and a race to be first on the beach. Most of us get on, but we're quite independent. There are a few people you learn to avoid, but I suppose that's just life.' According to Mike, there are about twenty regular professional collectors, but many more appear after a big landslide when the pickings are good. There is also an army of serious amateur collectors who are very dedicated and make some spectacular finds.

Since it gained World Heritage status, you need permission to dig into the cliffs and beach bedrock. There is an official code of conduct to protect the coastline and its fossils that encourages the reporting of important specimens and promotes the collection of fossils in a safe and sustainable way. In Mike's opinion it works well, and it's self-policing. 'People do report those who dig and at the moment it's fairly easy to apply for permission if you see something special, so it works.'

We search separately for a while, then Mike comes running back to me holding a brown lump. 'This is what you're looking for,' he says proudly. 'Poo!' I shout above the now pelting rain. 'But how do you know?' 'It's a different density to other rocks,' he explains, 'so it washes up in certain places, a bit like your things on the Thames, I suppose. If you look at the shape, you can tell it's poo, and those little black bits are the scales and bones of the fish it ate.' I am delighted.

'Shall we split some rocks now?' says Mike. He vanishes off and reappears with two large smooth grey nodules. To me they look like all the other rocks, but Mike's got the eye, and he knows where the treasure is hidden. He cleans off the mud, then rolls the stone expertly in his hands. 'Sometimes you see the keel of the ammonite sticking out around the edge,' he says, 'otherwise I need a good place to put the chisel where I won't damage any fossils near the surface. We should end up with the internal cast on one side and the shell on the other,' he says, holding the rock between his feet. Three perfectly positioned strikes of his hammer and it splits cleanly in two. There is a pause and I find myself holding my breath. Slowly, almost theatrically, he opens the nodule like a book to reveal three small spiral rock snails, opaque crystal cream against the dull grey rock.

Mike sighs. 'It's the first time they've seen the light in millions of years. Such beauty this rock's been hiding, and what a journey.' The last time these creatures saw daylight they were swimming in a shallow tropical sea, close to where the top of Africa is now. When they died, they drifted down into thick sediment and fell between branches and twigs that were already lying on the sea floor. As the plates moved and the Earth shifted, geology brought them to Lyme.

Wednesday 30 November (low tide 0.96 m @ London Bridge, 12.37)
Central London – North and South Bank

I leave home at 5.50 a.m. to a meagre dawn chorus. Now the evenings are so dark, most of my trips to the river will need to start early. The morning mist eventually burns away and the sky turns cold, clear and blue. I search for a while on my hands and knees next to a large crow, who is turning over stones and roof tiles looking for food. As I search, I mull over the padlock I'd found on my last visit. I have decided to leave it alone and not to risk destroying it by trying to knock off the rust. I think about it again and mentally confirm the decision is right. Eventually the crow stalks off to another patch and I look up to see a man in Lycra trying to push a racing bike over the shingle. He is trying to look as nonchalant as he can, as if he had always intended to bring his expensive bike down to the foreshore, but it is obvious he has taken a wrong turn.

Burnt, spent rockets from Bonfire Night are still washing up and so has a pumpkin whose carved face is disintegrating and folding in on itself. I kick it absent-mindedly and it smushes into pieces; I wonder if the crow will come back and eat it. Beside it are two bone ends for the bone tree and a yellow-plastic toy plane is emerging wing first from the mud. It has 'MADE IN HONG KONG' on its undercarriage and it looks like an old cereal-box toy; later research confirms it's been hanging around since the 1960s.

Then I spot a mudlark I recognise walking decisively towards me. 'Have you heard the news?' he says when he reaches me. I have, there aren't many people on the foreshore who haven't, and it's caused quite a stir. About a week ago the PLA announced that, although they were still renewing old permits, they were pausing the issue of new ones to 'protect the historical integrity of the Thames foreshore'. In 2018 there were around 200 people with mudlarking permits and now there are over 5,000, with numbers leaping during lockdown when people were looking for outside space and something to do. How many permits are 'live' and how many have only been used once or twice is anyone's guess, but the concern is that the PLA might go even further and ban mudlarking altogether.

'People are worried,' the mudlark says. 'I'm going over to the north side to see what they're saying, do you want to come?' But I don't want to get caught up in a group panic, so I politely decline and we part ways at the river stairs in front of the Globe theatre. I push it to the back of my mind to think about later and carry on as I'd planned. At the bottom of the river wall I pick up three semi-precious stone bracelets that look as if they were thrown in last night, perhaps as an offering to the river, then I race another mudlark across a bank of sharp rubble to my hunting ground east of Southwark Bridge, and I am glad I did. I had given up searching this area earlier in the year when a blanket of sand appeared almost overnight and smothered it, but it is being uncovered again and there is a new line of small metal and lead that's worth picking through.

I work my way along the waterline, then up onto the rubble to the bare patches of mud where objects sometimes catch. As I wobble across the bricks and flint, I see a rectangular thing, just smaller than the size of a matchbox, and recognise it straight away as a decorated pewter buckle plate. Colin later confirms it is fourteenth or fifteenth century and probably a cheaper version of the more upmarket copper alloy or silver buckle plates that were produced at the time.

The end of a leather belt would have slipped into the plate and been secured by a pin, now missing. The buckle frame, which is also missing, would have done the belt up. The decoration is Gothic in style and similar to designs found on furniture and church screens of the time. Colin promises to send me one of his shiny replicas, which is almost exactly the same but slightly smaller, for a three-quarter-inch belt instead of a one-inch belt like mine.

There is enough time left on the tide to cross Southwark Bridge for a quick look on the north side, which is quieter now some people have left, and I drop down a short ladder onto the foreshore at Queenhithe, where it is sunnier and warmer. The stinking stream of effluent that has been flowing from the Walbrook has finally dried up and a moorhen struggles past against the tide. A seagull swoops down and grabs a pink plastic dummy out of the strandline of plastic and looks at me comically. I wave my arms to try and make it drop it, but it flies off like a giant white-winged baby.

WINTER

CODE: 3112.22.D25

OBJECT:	19th imitation guinea token
MATERIAL:	Copper alloy
DATE FOUND:	31/12/2022
LOCATION:	Deptford
NOTES:	A pierced token promoting Millie Edwards and General Mite, the 'Royal American Midgets'. Imitation guinea tokens were mostly advertising gimmicks. Some could be exchanged for goods and services, and they were also used as gaming tokens. The bust of George III on the obverse avoided counterfeit laws, since most of these tokens were made in the reign of Queen Victoria.

December

Thursday 1 December
The Society of Antiquaries

According to the meteorological calendar, winter begins today. It is the end of the year, winter is back, time has passed, and another twelve months have ticked by. The moment we are born a clock starts ticking that only stops when we die. Our time here is made up of the present, the past and the future. The present is fleeting – catching it is like trapping a flea. Even as I write this, my present is becoming my past. I am writing my way into the future, briefly engaging with the present and creating the past. Past time is concrete, factual and real. The future is uncertain, unwritten and not yet real; it hasn't happened yet, so can it even really be called time or is time in general just an illusion? Mother Nature doesn't set her watch, birds don't roost according to the hands on a clock and flowers don't open on the hour; only humans choose to chain themselves to something as spurious and contrived as time.

Sarah says I have no concept of time, but I do. I think about it a lot. Maybe not immediate time, but I know where I stand in the wider concept of time, important time, the time before and after my fleeting visit to this world. I know how little of this great passage of time we occupy. I just lose track of unimportant everyday time sometimes, and I suppose I try to trap time through the things I keep in my Wunderkammer. Perhaps my search has become too obsessive and my research too feverish, but the tiny slivers of the past I find in the mud help me to understand my place. That I may just be a fragment, but I still fill a useful space in time.

Most people are searching for something; perhaps mudlarks are searching harder than others. I don't know if I will ever find what I am looking for because I don't think I really know what I've lost or what I'm searching for, or maybe I just don't want to find it. All I do know is that I can't stop looking.

I spend the day at the Society of Antiquaries, lost in their library and museum. So lost, in fact, that I lose track of time. I am half an hour late for the meeting where I am supposed to be formally admitted, and by the time I slide in at the back, the lecture has already started, and I have missed my official admittance as a Fellow of the society.

I sit at the back of the room on a hard wooden bench, half-listening to the career of the herald and antiquary Peter Le Neve (1661–1729). There is a wall of tweed between me and the speaker, and the room smells of wood, leather and dust. A large clock above the door ticks away the hour, rows of pan-faced Tudors gaze down at me from paintings on the walls and I can see the famous old black bicorn or cocked hat that is always placed in front of the president during meetings.

After the lecture, I sidle up to the society's president and apologise. 'Don't worry,' he says with a gentle smile. 'Come next time. Sherry? Dry, medium or sweet?' I take the last glass of dry and promise not to be late next time.

Friday 2 December (low tide 1.36 m @ North Woolwich, 14.47)

A Secret Location

Locations are a prickly topic within the mudlarking community. Some people are happy to divulge where they search, others are not. I've been criticised for being too open about locations, but it's not rocket science. There will be more stuff where the river was busiest in the past. Of course, I don't give away my best find spots within those locations, and they can change overnight at the whim of the river anyway, but telling someone to go to Rotherhithe to mudlark, for example, won't help them any more than telling them to go to London to buy a pair of jeans. The Rotherhithe foreshore is long and complex, and the only way to find anything is through time, patience and persistence.

There's only one location I keep completely secret and that's where there is a mini-hoard of sixteenth-century gold that has been gradually washing out of the mud. None of the pieces is longer than around 0.7 in and most are broken, incomplete or crushed, but enough is left of them to see that they were exquisitely crafted and would have originally belonged to someone of high social standing. Mudlarks have been finding single pieces spread across a fairly small area since around 2014. The Museum of London has recorded hundreds of them, including single links and tiny shavings, as well as the delicately decorated lace aglet I found several years ago. Most likely the gold was in a pocket or pouch – perhaps it was scrap gold for melting down – which means that there might be a rich epicentre waiting to surface one day.

There is already someone at the mini-hoard site today. She is checking all the usual places, and I can tell she knows exactly what she is doing. It is the golden elephant on the foreshore. Neither of us wants to mention it, so we do a little verbal dance: 'Found anything?' 'Not much, are you looking for anything special?' 'Just pebbles. I don't really know what I'm looking for.' Though she does, because I've seen her here many times before. I want to ask if she's found any gold, but I can't because that would be telling her that it's here, even though she probably already knows. And she's not the only one. I once met a man who gave me an exaggerated wink and said cryptically, 'I know what you're here for.'

Mudlarks do this all along the foreshore. A subtle exchange to suss out who's in the know and who isn't, what's been found and what hasn't. Some mudlarks lie: one man is always complaining that he's never found anything, but for someone who doesn't find much, he spends a lot of time looking. I've fibbed to people too. I've told them I found something in a completely different place from where I really found it and watched them scuttle off to see if I've missed anything else. It's a game, and I'm sure people have done the same thing to me.

At the hoard site, I pounce on every tiny white stone that shines in the weak winter sun and fragments of little curled snail shells that deceive me. I pick out an oyster shell that looks like the edge of a silver coin, and every humble, shiny pin looks like treasure. I search with my nose close to the ground, and I look in every nook and cranny, but there is no glint of gold today.

Sunday 11 December (low tide 1.13 m @ London Bridge, 09.57)
Central London – South Bank

London is covered in a thick swirling cloud of fog that seems to pour from the river itself. It formed when the land cooled quickly under last night's cold, clear sky and moisture in the air condensed. The tiny droplets of water are not large enough to land like rain, so they float en masse like a cloud. Even if they were to fall, there wouldn't be enough water to call it rain. If an Olympic swimming pool was filled with fog, it would yield just 1.25 litres of actual water.

Fog, mist and haze are defined by obscurity. The tipping point from mist to fog is a visibility of less than three-quarters of a mile. Today I can't see much more than 20 m in front of me. The bridges loom out of the whiteness and the opposite bank doesn't even exist. The river could just as well be an estuary or a ditch. The fog has thrown a white veil over the city and everything modern is gone. Even the short distance between my eyes and the foreshore is misty, the fog is curling around my boots, and what I can see of the water is flat and eerily still. Everything is suspended in a supernatural cloud that trickles into me through my ears, my nose and my mouth. It sinks onto my lungs, which feel thick and marshy, as if the river has crept inside me and settled there.

Years ago, the fog was even more sinister. Particles of nineteenth-century industrial pollution coloured it green, brown, yellow, grey, orange and sometimes even black with soot. It was thick and acrid. It swirled down streets and alleyways, blocked out the sun, consumed buildings, poured through keyholes and squeezed under doors. It lurks in the background of old photographs of the Thames, swallowing up ships, and casting a gauzy backdrop through which scant shadows and suggestions of life loom like apparitions.

The thick smogs of the early 1950s, nicknamed 'pea soupers', were laden with sulphur dioxide that smelled of eggs and turned them an evil shade of yellow. They directly and indirectly killed over 10,000 people and paralysed the city, bringing traffic, trains and the river to a standstill. The PLA issued long guide sticks to the captains of ships and barges so that they could find the edge of the docks in the smog. In theatres, it was sometimes so bad that the audience couldn't see the stage, and on the afternoon of 16 January 1955 it blotted out the light almost completely, creating an apocalyptic panic. As with the Great Stink a hundred years earlier, legislation was passed and slowly the air began to clean up, but even today a thin haze of smog still covers the city on most days, and the relatively clean white fogs that hang over the Thames are laden with lead, hydrocarbons and carbon monoxide.

Despite knowing this, I like fog. It is an extra layer against the big world, a duvet I can pull over my head and hide under. I feel safe tucked into a cold damp cloud, my small world just a few grey feet of foreshore and some muffled sounds: a jogger's rhythmic puffing, motorbikes crossing bridges and the sudden rustle of feathers. I look up and see three seagulls pass low enough to briefly enter my sanctuary before vanishing into the fog again.

My finds are simple: a seventeenth-century buckle and a black-glass button, probably sixteenth century. It is shaped like a Jelly Tot with a swirled back where the molten glass was wound into a blob. A rather uninspiring thick brown pot base with white inclusions in the clay may be Anglo-Saxon, but it may also be Roman. It's hard to tell the difference between the two. According to an archaeologist friend, apart from microscopic analysis, the only true test is to use acid. The inclusions in Anglo-Saxon pottery are usually shell, which fizzes, while Roman inclusions are mostly flint, which doesn't.

Unseen, and directly opposite me on the other side of the river, my Anglo Saxon-loving friend Flora has made an exciting find. I only find out when I check my messages on the way home, otherwise I could have crossed the river and seen it for myself. 'I've been wandering around dumbstruck all morning after this washed up,' she messages, and attaches a photograph of a small pewter arrow. Colin has already got back to her, and he is mystified. He hasn't seen anything like it, but he has a couple of theories. It is either a pilgrim badge for St Edmund or St Sebastian, both of whom were killed with arrows, or it is an amulet against the plague, since there was a medieval belief that disease was delivered to Earth on the tips of arrows. On a cheerier note, he also suggests it could be love's arrow, which would make it a medieval love token.

Thursday 15 December (low tide 1.15 m @ London Bridge, 11.49)
Central London – South Bank

I am wrapped in so many layers I can barely move my arms and my forefinger and thumb on my right hand are completely numb. I had thought it would be a good idea to cut the finger and thumb off my neoprene gloves, to make it easier to pick things up off the foreshore, but I'm regretting it now. However many times I clumsily click them together, they are not warming up.

I passed snow-covered fields on the drive up to London and the car dashboard registered an outside temperature of -5° C. On the river there is the added chill of a biting westerly wind that snips at my nose and cheeks, and it is already freezing me to the core. The sky is sharp and blue with a faint tinge of pink that threatens more snow to come, but cold has never kept me from the river. I have mudlarked as fat flakes of snow fell on my back with a scratchy puff and covered the foreshore around me, settling in a thin white layer closest to the wall and fizzing away by the water. But it's rare for snow to settle in central London, so seeing the Globe with a snow-covered thatch and negotiating an icy river path is a novelty.

Centuries ago, these kinds of temperatures were normal. The Little Ice Age began around 1300 and petered out towards 1850. Crops failed, cattle and sheep froze to death in fields, people starved and found it hard to find water that wasn't frozen solid. The wider, shallower river froze more easily too. The fat piers of the medieval bridge trapped lumps of floating ice and debris, creating a dam that slowed the river's passage to the point that it froze over completely, and great frost fairs were held on the captured ice.

In 1788–9 the river froze from Putney to Shadwell and a bear hunt was had on the ice at Rotherhithe. When the ice finally melted, and the treasures that had been dropped and lost in the muck and slush underfoot sank to the bottom of the Thames, the ice floated away in great chunks that dragged away a ship in Rotherhithe along with the riverside pub it was anchored to, killing five people who were asleep in their beds.

At Bankside, the river stairs are icy, and I descend them more carefully than usual. The swans have pressed the snow flat with their huge feet and their prints are preserved under a layer of clear ice. Safely at the bottom, I look back up the stairs and see Fiona looking back down at me. 'Be careful,' I shout, as she begins her slow descent to meet me.

Fiona is a foreshore archaeologist and we met years ago through the river. She lives by the river, and she has devoted much of the last thirty years of her life to studying it, culminating in a PhD which she completed as she was approaching sixty, after five of her six children had left home. She has long grey hair, an indomitable walk, and never wears boots on the foreshore, however cold it is, insisting that trainers are safer because they 'let the water flow out if they get wet'.

She steps off the stairs with a relieved huff and looks up at the dome of St Paul's, which is stark against the soft, pink, snow-laden sky. 'They found an almost complete pipe up there,' she says. 'They were doing repairs to it in the nineties, while they were building the Millennium Bridge. There was scaffolding all over it, right up to the top. Three of the men who were working on it were on one of my foreshore walks. They told me they'd found a complete one on a ledge on the little tower at the top of the big dome – it must have been left up there by another worker back in the day.'

'East or west?' I ask. 'Let's go west first,' she says. 'It's been a while since I've been here.' There are about four places Fiona visits regularly and Southwark is one of them. 'They need to be easy to access,' she once said to me. 'I don't drive and anything I collect goes in my backpack, so I don't want to walk far.' Over the years she's watched how the river works: how heavy rain upstream increases flow and affects the tides, and how tides coming in from the sea clash with river flow to increase erosion. With the shifting mud she's also seen mudlarks come and go and witnessed their effect on the foreshore. 'Because it's built up over so many years, it has a sort of locking system. If you dig a hole, you break the lock, even if it's backfilled. We've lost a lot of foreshore that way.'

Most of the good spots have already been peeled back by another mudlark, so I head for the waterline, where the foreshore is still untouched, and am rewarded almost immediately. A nineteenth-century clay pipe of good length is caught against a block of concrete – five seconds later and it would have been washed away by the next passing Clipper. It is decorated with a swan on each side and an acorn on the bottom. As I bend down to wash the mud out from inside it, another Clipper passes by, pulling the water out with it and revealing what looks like a bone knife handle. I have seconds to react before it is covered again, so I snatch it up and jump back out of the way of the wave that follows.

The thing in my hand is too heavy to be bone. I turn it over with frozen fingers and look at it carefully. It is the length of a knife handle, but it is made of metal, hollow at one end with a triangular point at the other. It looks like the metal tip from a crossbow bolt, a socketed quarrel, and if that's what it is, it would date to around the fourteenth century.

The word quarrel comes from the French *carreau*, meaning 'square', which is the perfect shape to punch through plate armour on the battlefield. If it hit flesh, the result was similar to being shot by a bullet, and modern tests have shown that just the impact was enough to kill by blunt trauma alone. The Hundred Years War raged through much of the fourteenth century, so I wonder if it had been destined for the battlefields of medieval France.

I find a very worn medieval floor tile nearby, which I give to Fiona for her collection, and she finds three pieces of carved stone that I hadn't noticed before. One looks like the square top, or bottom, of a column or pillar, another has a low arch cut into it, and a smaller piece is shaped and pitted with chisel marks. Fiona takes the smallest piece, and I hide the pillar bottom. Some things need to stay where they are.

'What's your best thing, Fiona?' I ask as I hand her the tile. 'Easy,' she says, 'it's a Mesolithic handaxe, a Thames pick. I found it close to here when I was recording prehistoric trees in the nineties.' It was an accidental find. Fiona was recording the drowned prehistoric trees that emerge at Southwark on very low tides. They can be seen all along the Thames and date from when sea levels rose around 3,750 years ago and drowned the scrub and woodlands that lined the prehistoric river. The remains of trees and bushes are preserved as humps of dark brown peat and perfectly preserved waterlogged branches, trunks and roots that melt away with the tide once they are exposed. Worked flints like Fiona's are evidence of the families that lived and hunted among the trees and scrub, the earliest Londoners.

'On one visit I dropped my pencil, bent down to pick it up and there, between my feet, was a tranchet axe, about four inches long.' *Tranchet* is French for 'slice', which describes the way the cutting edge was created by removing a large flake (or slice) to create a sharp edge. 'It had clearly been used,' she says, 'and it was unusual because it had a kind of twist that just didn't sit in the right hand, so it must have been a left-handed axe, and I love that!'

We pass under the Millennium Bridge, and I ask if she wants to carry on to Southwark Bridge, acutely aware of the rubble we have to pass over and her stiff knee, the result of an accident on an archaeological dig in Turkey a few years ago. 'I'll be fine,' she says with typical stoicism. We reach the bridge, and I leave her poking about in the rubble while I walk on a little further.

In a patch of metal, I spot the familiar shape of a hammered coin, small and dark grey with tarnish, a fish scale among the rust. I can just make out a cross on one side, which tells me it is a medieval penny, but once again the king's face flummoxes me. I'll need to ask Coin Man Paul, I think as I wobble back to Fiona across the rubble.

Fiona is looking tired and cold when I reach her. 'Tea?' I offer. 'Oh yes,' she says, 'I'm freezing.'

Wednesday 21 December
Winter Solstice

The sun goes down on the shortest day of the year at 15.53 GMT. The day lasts seven hours, forty-nine minutes and forty-two seconds in London, and at precisely 21.47 the sun reverses its north-south motion and for a moment it stands still: *sol* (sun) *sistere* (to stand still). The astronomical winter has begun, and we are now in midwinter, the deepest and darkest time of the year. But the sun is no longer ebbing, it is climbing, and from now on the days will get longer and the nights shorter. As more light inches in daily, I can finally begin to look forward to more time on the foreshore.

Tuesday 27 December (low tide 0.3 m @ Chelsea Bridge, 11.58)
Chelsea

I never usually mudlark at Chelsea, mainly because it's hard to get to from where I live, and also because I've never found as much as I have further east. It always seems like a bit of a waste of time to go that way, but today I decide to make an exception, to satisfy my curiosity about a missing type.

It isn't the first type that ended up in the river. The most famous is Doves type, the 'Thames type' that has become folklore among mudlarks. It was created and destroyed by Thomas James Cobden-Sanderson, bookbinder, printer and member of the Arts and Crafts Movement. Between August 1916 and January 1917 he threw one ton (500,000 pieces) of lead type into the Thames at Hammersmith, destroying the most important work of his life and preventing his former printing partner and adversary, Emery Walker, from ever using it again. Since I wrote about it in my first book, *Mudlarking: Lost and Found on the River Thames*, more has been found, mostly by metal detectorists I think, but still comparatively little has been recovered. I have an 'f' and a comma, which is enough for me, but now I'm ready to search for another Thames type.

Charles Ricketts, artist, author and printer, founded the Vale Press in 1896. Like Doves Press, it was founded in response to the revival of the art and craft of making books. While Cobden-Sanderson named his press and type after the Doves pub, a couple of doors down from his house on the river, Ricketts named his after the place he lived in Chelsea, just a few hundred yards from the river. They both created their type with meticulous care and obsession, and they both chose the Thames in which to dispose of it.

Ricketts created three fonts for his press: the Vale, the Avon and the King's, his favourite and the most experimental of the three. When he dissolved the press in 1904, Ricketts had the lead type melted down – it was too valuable to throw away – but he wrote: 'The punches and matrices are for the most part in the Thames, and on the completion of the last page of this pamphlet, the type becomes type metal again.' In 1937, the British Museum Print Room Acquisitions Register recorded the deposit of the matrices for all three Vale fonts, but these were later mislaid. So, from what Ricketts wrote, it suggests all that is left, the punches that were used to create casting moulds of the types, is lying at the bottom of the river.

The Vale, where Ricketts lived with his long-term companion and fellow artist Charles Shannon, has long since been demolished, but I find it on the Ordnance Survey map of 1895 and follow the most direct route on my computer screen from Ricketts' house to the river. Assuming he had thrown the punches into the water at Chelsea, it makes sense that he disposed of them somewhere between Albert Bridge and Chelsea Bridge. If he threw them off one of the bridges, there is no hope of ever finding them, but if he went to the riverside there is a wafer-slim chance of finding something.

The foreshore is predictably bare and covered with shingle. I'm late and arrive after low tide. I don't have much time, so I note the pinch points that are likely to cut me off from the stairs. I start at Albert Bridge and walk the full length to Chelsea Bridge along the water's edge. I find a broken nineteenth-century bottle, a lot of mainly blue-and-white Victorian crockery shards, the base of a ceramic Victorian toothpaste pot, a nineteenth-century clay pipe bowl and half a Georgian wig curler, which I wasn't expecting. I look wistfully across the river to the south side where it is said dredgings from Shadwell were dumped on the foreshore at Battersea, and wonder if I'd be doing better there.

I reach Chelsea Bridge and walk back along the wall. More broken crockery and some smashed stoneware lemonade or ginger-beer bottles. My last sweep has to be quick, as the tide is threatening to cut me off. I zig-zag all the way along the foreshore between the water's edge and the wall and again I find nothing of note. I am not surprised – there wouldn't be many punches and I suspect Ricketts did a much better job than Cobden-Sanderson at getting rid of them, but that's not to say they'll never turn up. Perhaps I'll come back again: maybe one day when the tide is ridiculously low or a boat has churned up the mud, a small letter cut into the end of a tall rectangle of metal will wash up. Meanwhile, it has been resurrected in this book; if you look at the headings for each month you will see Rickett's King's font. Cobden-Sanderson's Doves has been used on the section openers for each season.

Saturday 31 December (low tide 1.09m @ North Woolwich, 13.48)
Greenwich and Deptford

After the freezing weather of the last few weeks, it is curiously warm – 13° C – and if it wasn't raining in sheets, I wouldn't even need a coat. The trees I pass on my way into London are almost bare of leaves and their secrets have been revealed. Crows' nests and squirrels' drays balance in the top branches and below them green balls of mistletoe cling to thicker limbs. This evergreen plant grows hidden until this time of year, when the leaves of its host fall away to disclose a hidden magical parasite with no direct connection to the earth, a mysterious plant that's always fascinated me because it's too high up to be investigated properly.

Greenwich has the eerie, desolate New Year's Eve feeling I don't like. The wind is blowing a gale and the tall Christmas tree next to the *Cutty Sark* is waving precariously. Its lights twinkle in the gloom and reflect off the deserted, rain-slicked concrete around it. There is hardly anyone around, and the few people I pass aren't hanging about. Beside the river, the golden lights of an old-fashioned carousel are warm and inviting, the colours bright on such a miserable day, but there is no one riding the horses. It looks so much like part of a dream that I half expect it to fly away in a gust of wind. The mirrors flash and sparkle as it turns, and the organ music pipes out into the emptiness. It is the sound of summer, out of place on this grey, deserted midwinter day.

I am pleased to drop down onto the river, my familiar space where solitude is normal and friendly, and I pull up my hood against the driving rain. So much has changed since my summer visits. The good spot at the bottom of the stairs is covered in a layer of sand and shingle and a strange terrace of disturbed mud has appeared. It stands proud of the foreshore and has sharply defined edges, around fifty feet square. The mud in this patch looks turned over and stirred up, like Christmas cake, but instead of sultanas and nuts, it is studded with a muddle of shingle, mushed-up leather, pottery shards, smashed bones and pipe stems, most of which are broken into small pieces. It isn't fresh diggings, though. It looks as if the foreshore is revealing its old scars from when Greenwich was extensively dug over by mudlarks in the 1980s. There isn't much of this part of the foreshore that hasn't been disturbed, and whatever they missed would be eroding naturally now, but I wonder how much was damaged and broken.

My first find is half a narrow-waisted medieval left-footed shoe sole that's missing its heel. It is eroding out of a patch that looks undisturbed and old, a seam of gritty dark-grey mud, but this is deceptive. When I kneel down for a closer look, I see a large piece of white polystyrene deeply embedded in it, probably a quick fill by a digger years ago. I've seen all sorts emerge from old mudlarking holes: beer barrels, buckets, plastic bread trays, traffic cones, anything big enough to get the hole filled up quickly before the tide came in. Perhaps they had sliced the shoe sole in half with their spades too. I pick up some pins, a lead musket ball and quite a lot of Tudor pottery shards, and an hour off low tide I head for Deptford.

I pass a bunch of supermarket flowers tied to the railings; the cellophane is crackling in the wind. It isn't unusual to see them at this time of year, and flowers wash up on the foreshore too. It's a time of remembrance and family gatherings, of returns to the places that loved ones chose to be scattered, or just to their favourite place in life. Would I want to be scattered on the back of this cold grey basilisk? To become part of the mud? To mingle with leaves and silt, heavy metals and rust? To settle with the ghosts of the past among their clothes and shoes, cast-offs and losses, the oyster shells and bones of their meals, the broken crockery and glass of their arguments and accidents?

In the mouth of Deptford Creek, two swans are standing stoically in the rain and a handful of ducks are paddling about in the shallows. I tiptoe carefully into the mud and pick up a few pieces of delft; one is purple and white with the head of a bird that had been created almost 250 years ago with just a few strokes of the brush. It is a brief moment captured in time. There is a large piece of delicate eighteenth-century Chinese porcelain, and a cartouche from a seventeenth-century Bellarmine, though not enough to identify the town or city where it may have been made.

I climb back over the railings at the place where I had found leaf skeletons like the ones in my great-grandmother's Bible earlier in the year. New leaves are piled up against the river wall. They will be next year's skeletons. In a strandline of plastic below the leaves, two red Christmas-tree baubles sparkle among coffee cups, a broken coconut, a condom, three vapes and a medicine bottle. It is a depressing sight in the rain.

I follow the river path over Deptford Creek to Watergate Street. The big old cobbles are slippery with rain and at the bottom of the stairs there is a single forlorn gull sitting next to a heron, who is closely eyeing a puddle. He is exotic and primeval, thin and lonely, with tall, elegant legs and a smart grey morning coat. I almost expect to see a monocle balanced on the end of his long, sharp beak and a starched white collar around his neck. I say 'he' because all herons look like men to me – old-fashioned, smart-suited clerks. He's probably come off the creek, but I'm not sure he has chosen a very good fishing spot. The river is too deep beyond the revetment and the puddle isn't very big, but perhaps he isn't fishing at all. Maybe he has already eaten, his belly is full of eel and he is simply standing still for a few hours, as they do, to digest it.

The wind suddenly whips up a water devil between some small sailing boats moored just offshore. It spins a short spout of water up into the air and suddenly lets go, scattering it across the surface. I look up at the foreshore and I'm surprised at how busy it is. A man with a large dog is walking towards me and another man is jogging away from me. In the near distance, I see a metal detectorist in a green anorak, patiently sweeping his machine over the same spot again and again. He looks up and straight at me. Our eyes lock for a moment through the sheets of rain, then he turns and begins to walk in the direction of my shoe dump.

I beat him to it, but when I get there the tide isn't as low as it was on my last visit and there are no nails and little bits of metal to sift through. There are lots of pieces of shoe though and that is what I am looking for. Thick studded soles and heels are hanging out at all angles, and thinner pieces of upper leather flap in the wind. I kneel down for a closer look, and tug gently at one. It is stuck fast in the compacted mud. I try another, shredding my gloves on the hobnails, but it isn't moving either. Only one studded heel comes free easily, and half a large, beautifully patched and mended sole from a right-footed shoe.

They look like home repairs, not professional repairs by a cobbler. The little white dots around the edge are where the original linen thread had held the sole to the upper. It would have been treated with coad, a mixture of pine resin and beeswax, to stop the linen thread from rotting. Shoemakers often used the bristles from a boar's nose instead of needles, which they attached to the thread. They were cheaper than needles and followed the awl's path through the leather more easily, without getting caught on an eye.

The larger white dots around the repairs are the remains of tiny wooden pegs, about the size of a match, that were used like nails. I've found heels held together with wooden pegs before, which worked well in our clement climes, but not so in America. Prior to 1872 the soles of American Army boots were attached to the uppers by wooden pegs, which shrank and fell out in the dry western climate. They were replaced with brass screws that conducted the cold in winter and eventually linen thread.

I slip them both into my waist bag, then I see the familiar shape of a coin wedged tightly underneath one of the shoe soles. I pick it out with my fingernails. 'WITH MILLIE EDWARDS LOVE' reads one side, and on the other 'WITH GENERAL MITES COMPLIMENTS'. It is a nineteenth-century token made to look like an eighteenth-century guinea, with George III's bust on Millie Edwards' side. Some were made as gaming tokens, others for advertising and many (like this one) were pierced and sold as trinkets. It is a publicity token celebrating the union of two of the smallest people that ever lived: General Mite, Francis Joseph Flynn, who was born in New York in 1864 and is said to have never grown beyond twenty-two inches tall, and Lancashire-born Millie Edwards, who was said to be just nineteen inches high. They married and performed together as the Royal American Midgets, until Flynn died in Australia at the age of thirty-four. Millie continued to perform as Mrs General Mite and died in New Zealand in 1919, aged forty-two.

By 2.30 p.m. the weather and the light defeat me. The tide has turned, and I decide to call it a year. I walk back past the detectorist, who doesn't look up from what he is doing, to the stairs where three young men have just arrived and are talking excitedly about what they might find. I look back at Greenwich and see the bright lights of the carousel, still empty, still turning. The sailing boats have turned on their moorings and the river is coming back in again; later it will go out for one last low tide of the year. The river counts its age by tides and not years, and it cannot turn back. Like us, it has to move forward, into the unknown.

A List of Time Periods

Precise dates can vary, depending on who you're asking and which book you're reading.

Mesolithic (hunter-gatherers): 10,000–4,000 BCE

Neolithic (the first farmers): 4,000–2,300 BCE

Bronze Age: 2,300–800 BCE

Iron Age: 800 BCE – 43 CE

Roman: 43–410 CE

Saxon and Viking: 410–1066

Medieval: 1066–1485

Tudor: 1485–1603

Stuart: 1603–1714

Georgian: 1714–1837

Victorian: 1837–1901

Edwardian and the world wars: 1901–1945

Modern: 1945–present

Select Bibliography and Other Sources

PUBLICATIONS

Books and publications are listed once, under the chapter where they are first used.

January

Peter Ackroyd, *Thames, Sacred River* (London, 2007)

Ivor Noël Hume, *If These Pots Could Talk: Collecting 2,000 Years of British Household Pottery* (Milwaukee, 2001)

Henry Humpherus, *The History of the Origin and Progress of the Company of Watermen and Lightermen of the River Thames, 1515–1859* (three vols, 1874–86)

Marion Johnson, 'The Cowrie Currencies of West Africa', *The Journal of African History*, vol. 11, no. 1 (Cambridge, 1970)

Henry Mayhew, *London Labour and the London Poor* (London, 1851)

Ian Mortimer, *The Time Traveller's Guide to Regency Britain* (London, 2020)

Liza Picard, *Dr Johnson's London: Everyday Life in London 1740–1770* (London, 2000)

—, *Elizabeth's London: Everyday life in Elizabethan London* (London, 2003)

George R. Sims (ed.), *Living London: Its Work and Its Play, Its Humour and Its Pathos, Its Sights and Its Scenes*, vol. 1 (London, 1901)

Julia Smith, *Walks for Each Season: 26 Great Days Out in the Countryside Near London* (London, 2021)

February

Patrick Colquhoun, *A Treatise on the Police of the Metropolis; Containing a Detail of the Various Crimes and Misdemeanors By Which Public and Private Property and Security are, at Present, Injured and Endangered: and Suggesting Remedies for their Prevention* (London, 1796)

Daniel Defoe, *The Storm* (London, 1704)

Ivor Noël Hume, *A Passion for the Past: The Odyssey of a Transatlantic Archaeologist* (Virginia, 2010)

—, *Treasure in the Thames* (London, 1956)

Sarah Inskip and Angela Joy Muir, 'Material encounters: the alternative use of clay tobacco pipes in England and Wales, *c.*1600–1900', *Historical Research*, vol. 96, issue 272 (May 2023)

Ian Mortimer, *The Time Traveller's Guide to Restoration Britain, 1660–1700* (London, 2017)

Brian Read, *Metal Buttons: c. 900 BC – c. AD 1700* (Somerset, 2005)

Charles Roach Smith, *Catalogue of the Museum of London Antiquities* (London, 1854)

'Trial of Edward Goynes, alias Joynes', September 1739 (t17390906-6). Old Bailey Proceedings Online. www.oldbaileyonline.org

March

Clive Aslet, *The Story of Greenwich* (London, 1999)

Tracey Boorman, *The Private Lives of the Tudors* (London, 2016)

M. Channing Linthicum, *Costume in the Drama of Shakespeare and his Contemporaries* (Oxford, 1936)

Emily Cockayne, *Rummage: A History of the Things We Have Reused, Recycled and Refused to Let Go* (London, 2020)

Charles Dickens, *Bleak House* (London, 1852)

Nicholas Eastaugh, V. Walsh, Tracey Chaplin and Ruth Siddall, *Pigment Compendium: A Dictionary and Optical Microscopy of Historic Pigments* (London, 2008)

Sean Shesgreen, *Hawkers, Beggars and Quacks: Portraits from the Cries of London* (Oxford, 2021)

Brian Spencer, *Pilgrim Souvenirs and Secular Badges: Medieval Finds from Excavations in London* (London, 2010)

Ross Whitehead, *Buckles 1250–1800* (Essex, 1996)

April

Peter Barber, *London: A History in Maps* (London, 2012)

Stephen Croad, *The Thames Through Time: A Liquid History* (London, 2016)

Chris Ellmers and Alex Werner, *London's Lost Riverscape: A Photographic Panorama* (London, 1988)

Francis Grew and Margrethe de Neergaard, *Shoes and Pattens* (London, 1988)

James Stewart, *Plocacosmos, or the Whole Art of Hair Dressing* (London, 1782)

John Stow, *A Survey of London* (London, 1598)

June Swan, *Shoes* (London, 1982)

May

Geoff Egan, 'Treasure in the Thames', *London Archaeologist* 3 (1977)

Francesca Greenoak, *British Birds: Their Folklore, Names and Literature* (London, 1997)

Alex Wright, *The Bellarmine and Other German Stoneware* (Norfolk, 2009)

June

Peter Jackson, 'London is Stranger than Fiction', *Evening News* (12 January 1948)

Ralph Merrifield, 'Letters to the Editor: Mudlarking', *The Times* (26 July 1976)

Pharmacopoeia Londinensis, Royal College of Physicians (London, 1788 edition)

Gillian Tindall, *The House by the Thames and the People Who Lived There* (London, 2006)

Walter Thornbury and Edward Walford, *Old and New London: A Narrative of its History, its People, and its Places* (London, 1878)

Harvey Wickes Felter and John Uri-Lloyd, *King's American Dispensatory* (Cincinnati, 1898)

July

Lesley Brown (ed.), *The New Shorter Oxford English Dictionary on Historical Principles* (Oxford, 1993)

Stuart F. Elton, *Cloth Seals: An Illustrated Guide to the Identification of Lead Seals Attached to Cloth* (Oxford, 2017)

Ruth Goodman, *The Domestic Revolution* (London, 2020)

August

Robert Halliday, 'The Billy and Charley Forgeries', *London Archaeologist* (Winter 1986)

Frederick Marryat, *Poor Jack* (London, 1840)

Werner Muensterberger, *Collecting: An Unruly Passion: Psychological Perspectives* (Princeton, 1994)

September

William Glenton, *Tony's Room: The Secret Love Story of Princess Margaret* (London, 1965)

Henry Lea, *Illustrated History of the Great Fire, and A Biography* London, (London, 1861)

Liza Picard, *Restoration London: Everyday Life in London 1660–1670* (London, 1997)

George R. Sims (ed.), *Living London: Its Work and Its Play, Its Humour and Its Pathos, Its Sights and Its Scenes* (London, 1901)

Peter Webster, *Roman Samian Pottery in Britain*, Council for British Archaeology (York, 1996)

Lisa Yeomans, 'The Shifting Use of Animal Carcasses in Medieval and Post-medieval London' in Aleksander Pluskowski (ed.), *Breaking and Shaping Beastly Bodies: Animals as Material Culture in the Middle Ages* (Oxford, 2007)

October

Mike Dash, 'The Commoner who Salvaged a King's Ransom', *Smithsonian Magazine* (19 August 2013)

Christopher J. Duffin, 'Herbert Toms (1874–1940), Witch Stones and "Porosphaera" Beads', *Folklore*, vol. 122, no. 1 (April 2011)

Illustrated, 'They Find History', Joan Higgens (13 March 1948)

Kirby's Wonderful and Eccentric Museum; or Magazine of Remarkable Characters, vol. III (London, 1820)

Dorothy Menpes, *World's Children* (London, 1904)

D. O'Hara, *Courtship and Constraint: Rethinking the Making of Marriage in Tudor England* (Manchester, 2000)

Hugh Pope, 'For Shakespeare's Globe Theatre, Hazelnuts Supply the Metaphor', *Wall Street Journal* (12 November 1997)

Peter Richey, 'Mudlarking for Pleasure or Profit', *The Times* (12 August 1972)

Tiffany Stern, 'Time for Shakespeare: Hourglasses, Sundials, Clocks, and Early Modern Theatre', *Journal of the British Academy*, Shakespeare Lecture (read 21 May 2014)

Rebecca Struthers, *Hands of Time: A Watchmaker's History of Time* (London, 2023)

November

E. Eckstein and J. & G. Firkins, *Gentlemen's Dress Accessories* (Buckinghamshire, 2000)

Our Troubled Rivers, BBC2 (first aired March 2023)

Richard Rowe, *Episodes in an Obscure Life* (London, 1871)

Brian Spencer, *Salisbury and South Wiltshire Museum Medieval Catalogue, part 2: Pilgrim Souvenirs and Secular Badges* (Salisbury, 1990)

December

John Evelyn, *The Diary of John Evelyn* (London, 1818)

Brian Fagan, *The Little Ice Age: How Climate Made History 1300–1850* (London, 2019)

Richard Hingley, *Londinium – A Biography: Roman London from its Origins to the Fifth Century* (London, 2018)

Douglas C. McChristian, *Regular Army O!: Soldiering on the Western Frontier, 1865–1891* (Oklahoma, 2017)

Charles Ricketts, *A Bibliography of the Books Issued by Hacon & Ricketts* (London, 1904)

Walter Thornbury, *Old and New London: Volume 3* (London, 1878)

OTHER USEFUL SOURCES

Websites

Agas Map: www.mapoflondon.uvic.ca

Bag seals: www.bagseals.org

Blake Society: www.blakesociety.org

Bodleian Libraries: www.bodleian.ox.ac.uk

British History Online: www.british-history.ac.uk

Brunel Museum: www.thebrunelmuseum.com

Burlington House: www. burlingtonhouse.org

Charmouth Heritage Coast Centre: www.charmouth.org

Chipstone Foundation: www.chipstone.org

Company of Watermen and Lightermen: www.watermenscompany.com

Currency Converter: www.nationalarchives.gov.uk/currency-converter

Diary of Samuel Pepys: www.pepysdiary.com

Doves Type: www.typespec.co.uk/doves-type/

Encyclopedia Britannica: www.britannica.com

English Heritage: www.english-heritage.org.uk

Erica Weiner: www.ericaweiner.com

Geological Society: www.geolsoc.org.uk

The Glossop Cabinet of Curiosities (Tim Campbell-Green): www.glossopcuriosities.co.uk

Historic England: www.historicengland.org.uk

John Rocque Map: www.locatinglondon.org

Layton Collection: www.thomaslayton.org.uk

Leaden Token Telegraph: thetokensociety.org.uk/ltt/

Lionheart Replicas: www.lionheartreplicas.co.uk

London Fire Brigade: www.london-fire.gov.uk

Lyme Regis Museum: www.lymeregismuseum.co.uk

Mary Rose Trust: www.maryrose.org

Met Office: www.metoffice.gov.uk
Mr Pepys' Small Change: www.c17thlondontokens.com
Old Bailey Proceedings Online: www.oldbaileyonline.org
Port of London Authority: www.pla.co.uk
Portable Antiquities Scheme: www.finds.org.uk
Royal Artillery Museum: www.royalartillerymuseum.com
Royal College of Physicians: www.rcplondon.ac.uk
Royal Family: www.royal.uk
Royal Mint: www.royalmint.com
Royal National Lifeboat Institute: www.rnli.org
Royal Society for the Protection of Birds: www.rspb.co.uk
Royal Swan Upping: www.royalswan.co.uk
Shakespeare Birthplace Trust: Shakespeare.org.uk
Society for Clay Pipe Research: www.scpr.co
Society of Antiquaries of London: www.sal.org.uk
Southwark Cathedral: www.cathedral.southwark.anglican.org
St Magnus the Martyr: www.www.stmagnusmartyr.org.uk
St Paul's Cathedral: www.stpauls.co.uk
Tanners Company of Bermondsey: www.tannersofbermondsey.org
Thames 21: www.thames21.org.uk
Tideway: www.tideway.london.com
UK Parliament: www.parliament.uk
Visscher map: www.panoramaofthethames.com
Woodland Trust: www.woodlandtrust.org.uk
World Gold Council: www.gold.org
Worshipful Company of Cutlers: www.cutlerslondon.co.uk
Worshipful Guild of Apothecaries: www.apothecaries.org

Museums

British Museum, London: www.britishmuseum.org
Brunel Museum, London: www.thebrunelmuseum.com
Charmouth Heritage Coast Centre, Dorset: www.charmouth.org
Dennis Severs' House, London: www.dennissevershouse.co.uk
Fitzwilliam Museum, Cambridge: www.fitzwilliam.com.ac.uk
Foundling Museum, London: www.foundlingmuseum.org
Globe Exhibition, London: https://www.museumslondon.org/mus
   eum/23/shakespeares-globe-exhibition

Imperial War Museum, London: www.iwm.org.uk

Jorvik Viking Centre, York: www.jorvikvikingcentre.co.uk

Little Woodham Living History Village, Gosport, Hampshire: www.littlewoodham.org.uk

London Mithraeum (Temple of Mithras): www.londonmithraeum.com

Lyme Regis Museum, Dorset: www.lymeregismuseum.co.uk

Mary Rose Museum, Portsmouth: www.maryrose.org

Museum of London: www.museumoflondon.org.uk

Museum of the Home, London: www.museumofthehome.org.uk

National Museum of Scotland, Edinburgh: www.nms.ac.uk

Pitt Rivers Museum, Oxford: www.prm.ox.ac.uk

Royal Armouries Museum, Leeds: www.royalarmouries.org

Royal Artillery Museum, London: www.royalartillerymuseum.com

Royal Museums Greenwich, London: www.rmg.co.uk

Royal Palaces: www.hrp.org.uk

Shakespeare Birthplace Trust, Stratford-upon-Avon: www.shakespeare.org.uk

St Fagans National Museum of History, Cardiff, Wales: www.museum.wales/stfagans/

Thames River Police Museum, London: www.thamespolicemuseum.org.uk/museum.html

Victoria and Albert Museum, London: www.vam.ac.uk

Vindolanda, Hexham, Northumberland: www.vindolanda.com

Weald and Downland Living Museum, Singleton, West Sussex: www.wealddown.co.uk

Acknowledgements

I've met so many people over the years who have been generous with their knowledge, too many to thank individually but I hope they know who they are.

Adrian Green (Director of Salisbury Museum), Alex Wright (stoneware specialist and proprietor of Antiques of London St, Swaffham, Norfolk), Andrew Honey (Book Conservator at the Bodleian Library), Andrew Nunn (Dean Emeritus of Southwark Cathedral), Charles Brooking (architectural historian), Colin Torode (pilgrim badge specialist at www.lionheartreplicas.co.uk), the Department of Archaeology and Conservation at Cardiff University, Dr David Higgins (clay pipe specialist), Dr Francesca Galligan (Assistant Librarian, Rare Books at the Bodleian Library), Emily Naish (archivist at Salisbury Cathedral), Erica Weiner, Emma Bridgewater (for a bed and the bittern), Dr Fiona Haughey (foreshore archaeologist), Flora Spiegel (mudlark), Graham de Heaume (mudlark), Ian Richardson (Senior Treasure Registrar at the British Museum), Jason Whittaker and Steven Pritchard (the Blake Society), Jo Ahmet (archaeological finds specialist and former Finds Liaison Officer for Kent), Johnny Mudlark and Liz Friend, Jon Dollin (Head of Visitor Engagement at Southwark Cathedral), Josh Maiklem, Julia Smith (friend, mudlark and author of her own wonderful book of walks @jswalks on Instagram), Kate Bagnall (Society of Antiquaries of London, Museum Collections Manager), Lloyd de Beer (Curator of Medieval Britain and Europe at the British Museum), Michael Rawson (Sub Dean and Canon Pastor at Southwark Cathedral), Mike Harrisson (fossil hunter), Mike Pitts (for helping me to become a Fellow), Mike Webber, as always, Paul Buck (former Curator of Horology at the British Museum), Paul Gray (my coin man), the Port of London Authority (Jim Trimmer, Nick Tenant and Pippa Barber), Professor Michael Lewis (Head of the Portable Antiquities Scheme at the British Museum), Dr Rebecca Struthers (horologist), Robert McPherson (ceramics specialist), Sam Caethoven (for helping me with the fossils), Sean Clarke (mudlark), Simon Hurst, Stuart Wyatt (Finds Liaison Office for London), all the Maiklems in Cornwall, the Very Reverend Nicholas Papadopulos DL (Dean of Salisbury), The Worshipful Company of Clockmakers, Tina Schinabeck (Registrar at the Chipstone Foundation), Tracey Boorman.

Thank you to everyone at Bloomsbury for another wonderfully professional and friendly publishing experience, especially Alexis Kirschbaum for believing in me a second time, my editor Anna Vaux for simply being the best person to work with again, Ariel Pakier for her brilliant editorial guidance, Juliet Brooke for seamlessly picking up from Ariel, Lauren Whybrow for coping with my obsessive detailing, Shanika Hyslop for delightfully organising me, Carmen R. Balit for my beautiful jacket, Kate Quarry, and the publicity and marketing team who are second to none. Thank you to Chiz Harward, man of the woods, for the beautiful illustrations, and huge thanks as always for the help and support of my fabulous agent

Sarah Ballard at C&W, Liv Bignold, and Eli Keren (I miss you).

Finally, the biggest thanks of all go to my family. To my mother for teaching me the precious art of looking. To the twins, my little life changers, for putting up with me through another book, and of course to my ever-patient Sarah, giver of the greatest gifts, without whom none of this would have been possible.

For everyone's kindness, love and support, thank you.

A Note on the Type

Most of the text of this book is set in Linotype Sabon, a typeface named after the type founder, Jacques Sabon. It was designed by Jan Tschichold and jointly developed by Linotype, Monotype and Stempel in response to a need for a typeface to be available in identical form for mechanical hot metal composition and hand composition using foundry type.

Tschichold based his design for Sabon roman on a font engraved by Garamond, and Sabon italic on a font by Granjon. It was first used in 1966 and has proved an enduring modern classic.

In homage to the two lost Thames types, Doves font is used for the title page and the section openers for the seasons. Charles Rickett's favourite from the Vale font family, Kings, is used for the headings of each month. The story of the Vale fonts can be read in the last chapter, 'December', and the full story of Doves is told in *Mudlarking: Lost and Found on the River Thames*.